IMPA Monographs

Volume 3

This series, jointly established by IMPA and Springer, publishes advanced monographs giving authoritative accounts of current research in any field of mathematics, with emphasis on those fields that are closer to the areas currently supported at IMPA. The series gives well-written presentations of the "state-of-the-art" in fields of mathematical research and pointers to future directions of research.

More information about this series at http://www.springer.com/series/13501

Jonas Gomes • Luiz Velho

From Fourier Analysis to Wavelets

Jonas Gomes
BOZANO Investimentos
Offices Shopping Leblon
Rio de Janeiro, Brazil

Luiz Velho
Instituto de Matemática Pura e Aplicada
Rio de Janeiro, Brazil

IMPA Monographs
ISBN 978-3-319-22074-1 ISBN 978-3-319-22075-8 (eBook)
DOI 10.1007/978-3-319-22075-8

Library of Congress Control Number: 2015945959

Mathematics Subject Classification (2010): 42-XX, 42C40, 65-XX, 65T60

Springer Cham Heidelberg New York Dordrecht London

Printed on acid-free paper

Springer International Publishing AG Switzerland is part of Springer Science+Business Media (www.springer.com)

To my family, the true force that propelled my carrier.

— Jonas Gomes

To my beloved daughter Alice Velho Chekroun and my grandson Levi Yitzchak Chekroun

— Luiz Velho.

Preface

From a digital viewpoint, most real-world applications can be reduced to the problem of function representation and reconstruction. These two problems are closely related to synthesis and analysis of functions. The Fourier transform is the classical tool used to solve them. More recently, wavelets have entered the arena providing more robust and flexible solutions to discretize and reconstruct functions.

Starting from Fourier analysis, the course guides the audience to acquire an understanding of the basic ideas and techniques behind the wavelets. We start by introducing the basic concepts of function spaces and operators, both from the continuous and discrete viewpoints. We introduce the Fourier and Window Fourier Transform, the classical tools for function analysis in the frequency domain, and we use them as a guide to arrive at the Wavelet transform. The fundamental aspects of multiresolution representation and its importance to function discretization and to the construction of wavelets are also discussed.

Emphasis will be given on ideas and intuition, avoiding most of the mathematical machinery, which are usually involved in the study of wavelets. Because of this, the book demands from the readers only a basic knowledge of linear algebra, calculus, and some familiarity with complex analysis. Basic knowledge of signal and image processing would be desirable.

This monograph originated from the course notes of a very successful tutorial given by the authors during the years of 1998 and 1999 at ACM SIGGRAPH - the International Conference and Exhibition on Computer Graphics and Interactive Techniques.

The notes in English for the SIGGRAPH course have been based on a set of notes in Portuguese that we wrote for a wavelet course on the Brazilian Mathematical Colloquium in 1997 at IMPA, Rio de Janeiro. We wish to thank Siome Goldenstein who collaborated with us to produce the Portuguese notes.

Rio de Janeiro, Brazil
May 2015

Jonas Gomes
Luiz Velho

Contents

Chapter 1
Introduction

In this chapter we give a general overview of the area of Computational Mathematics and computer graphics, introducing the concepts which motivate the study of wavelets.

1.1 Computational Mathematics

Mathematical modeling studies phenomena of the physical universe using mathematical concepts. These concepts allow us to abstract from the physical problems and use the mathematical tools to obtain a better understanding of the different phenomena of our universe.

The advance of computing technology (both hardware and software) has enabled the use of mathematical modeling to make simulations on the computer (synthetic simulations). These simulations allow for a great flexibility: We can advance in time, accelerate processes, introduce local and global external factors, and change the different parameters on the mathematical models used. These conditions of simulation in general are very difficult to be attained in real-world simulations.

Computational Mathematics is the combination of mathematical modeling with computer simulations. Computational mathematics represents a return to the birth of mathematics, where mathematical problems were directly related to the solutions of practical problems of our everyday life.

In this context, computer graphics provides different techniques to visualize the results from the simulation. This visualization enables a direct interaction of the users with the different parameters of the simulation and also enables them to collect valuable qualitative and quantitative information about the simulation process.

J. Gomes, L. Velho, *From Fourier Analysis to Wavelets*,
IMPA Monographs 3, DOI 10.1007/978-3-319-22075-8_1

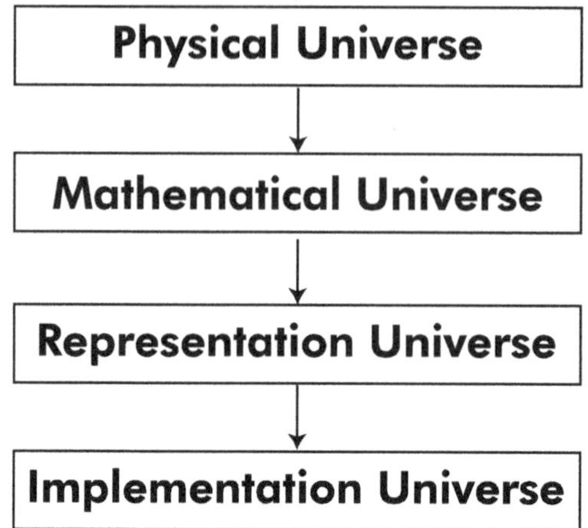

Fig. 1.1 Abstraction levels for computational mathematics

1.1.1 Abstraction Levels

A useful way to organize the ideas discussed in the previous section, which will be used in the book, consists in structuring the problem of computational mathematics using abstraction levels. These abstraction levels encapsulate the different phases of the simulation processing providing a better understanding of the whole process. We will use the paradigms of the four universes. This paradigm uses four levels: Physical universe, Mathematical Universe, Representation Universe, and Implementation Universe (see Fig. 1.1).

Level 1: Physical Universe In this level the problems are originally posed and correctly formulated. The elements of these abstraction levels consist of the different phenomena from the physical universe.

Level 2: Mathematical Universe This level contains tools from different areas of mathematics (Analysis, Geometry, Differential Equations, Algebra, etc.). The process of *mathematical modeling* consists in the association of elements from the physical universe (Level 1) with elements of the mathematical universe.

Level 3: Representation Universe To perform synthetic simulations the mathematical models must be discretized and implemented on the computer. The representation universe consists of the different elements of discrete mathematics and the mathematical methods of discretization. As part of this universe, we have virtual machines where we can describe algorithms. The discretization of a problem consists in relating the mathematical model associated with the problem in the mathematical universe (Level 2) to a discretized model in the representation universe.

Level 4: Implementation Universe This abstraction level is considered so as to allow for a clear separation between the discretization problem in Level 3 and the problems inherent to implementation. The elements of this universe consist in data structures and programming languages with a well-defined syntax and semantics. The implementation of a problem consists in associating discrete structures from the representation universe with elements of the implementation universe.

These four abstraction levels will be used along the book.They split the problems into four major sub-problems and favors an encapsulation of each of the sub-problems: At each level the problem is studied with a clear separation from its intrinsic issues with issues related to the other abstraction levels.

We should point out that this abstraction paradigm constitutes a characterization of computational mathematics. Below we will give some examples.

Example 1 (Measurements). Consider the problem of measuring objects in the physical universe. For a given object we must associate with it a number which represents its length, area, or volume. In order to achieve this we must introduce a standard unit of measure which is compared with the objects to provide a measure.

From the point of view of the mathematical universe, to each measure we associate a real number. Rational numbers correspond to commensurable objects, and irrational numbers correspond to incommensurable ones.

To represent the measurements, we must look for a discretization of the real numbers. A widely used choice is given by the *floating point representation*. Note that in this representation the set of real numbers is discretized by a finite set of rational numbers. In particular, this implies that in the representation universe we do not have the concept of incommensurable measurements.

An implementation of the real numbers using floating point arithmetic can be done using a standard IEEE specification. A good reference for these topics is [29].

The above example, although simple, constitutes the fundamental problem in the study of computational mathematics. In particular, we should remark the loss of information when we pass from the mathematical to the representation universe: incommensurable measures disappear. In general the step from the mathematical to the representation universe forces a loss of information, and this is one of the most delicate problems we have to face in computational mathematics.

Example 2 (Rigid Body Motion). Consider now the well-known problem of studying the motion of a small rigid body which is thrown into the space with a given initial velocity. The problem consists in studying the trajectory of the body.

In the mathematical universe we will make a simplification to model the problem. The rigid body will be considered to be a particle, that is a point of mass. This abstraction allows us to neglect the air resistance. Simple calculations well known from elementary physical courses allow us to solve the problem of determining the trajectory: a parabola

$$f(t) = at^2 + bt + c \; .$$

To visualize the trajectory in the computer we must obtain a discretization of the parabola. This discretization can be obtained by taking a uniform partition $0 = t_0 < t_1 < \cdots < t_n = t$ of the time interval $[0, t]$. Using this partition the parabola is represented by a finite set of points of the plane:

$$(t_0, f(t_0)), (t_1, f(t_1)), \ldots, (t_n, f(t_n)) .$$

These points constitute *samples* of the function that defines the parabola.

Finally, data structures to implement the samples are well known and can be easily implemented.

Note that in order to plot the parabola we must "reconstruct" it from the samples. Interpolation methods must be used in the reconstruction process. Again, the representation process by sampling loses information about the parabola. Depending on the interpolation method used we will not be able to reconstruct the parabola exactly, but only an approximation of it.

An important remark should be done. The solution to the problem in the above example was initially attained in the mathematical universe; it was an analytical solution providing the equation of the trajectory. The discretization of the analytical solution was done in order to visualize it. Another option would be discretizing the problem before solving it by using discrete mathematical models to formulate the problem [27]. Which is the best strategy? Discretize the problem to obtain a solution or compute an analytical solution and discretize it.

The answer to this problem is not simple. Initially we should remark that there are several situations where we do not have an analytic solution to the problem in the mathematical universe. When this happens discretization a priori to determine a solution is of fundamental importance. This is the case, for instance, in several problems which are modeled using differential equations: The discretization allows us the use of numerical methods to solve an equation where an analytical solution is difficult or even impossible to be achieved.

When the problem has an analytical solution in the mathematical universe we have the option to compute it before discretization. Nevertheless we should point out that the process of discretization should not be decorrelated from the formulation of the problem in the physical universe. In example 2, for instance, the parameter of the parabola represents the time, and we know that the horizontal motion is uniform. Therefore, a uniform discretization of time is the most adequate to playback the animation of the particle motion.

1.2 Relation Between the Abstraction Levels

We should not forget the importance of establishing for each problem the relation between the four universes in our abstraction paradigm described above. In its essence these relations enable us to answer the question: How close a computational

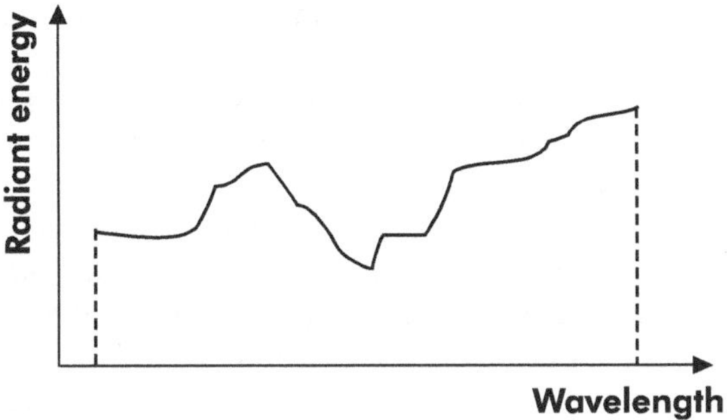

Fig. 1.2 Spectral distribution function

simulation is from reality? The answer to this problem depends on the relation between the different abstraction levels:

- Relation between the implementation of the problem and its discretization;
- Relation between the discretization and the mathematical formulation of the problem;
- Relation between the mathematical model of the problem and the original problem in the physical universe.

The relationship between the mathematical and the representation universe is of most importance. We have seen that there are losses in the representation process, and a very important issue consists in recovering the continuous model from the discrete representation. This step is called *reconstruction*. Since the representation is a loss process, reconstruction in general is not possible to be attained exactly. The example below will make this point clear.

Example 3 (Representation of Color). Color is produced by an electromagnetic radiation in the visible band of the spectrum. Therefore from the point of view of the mathematical universe, a color is determined by its *spectral distribution function*. This function associates with each wavelength λ the value of the associated radiant energy (see Fig. 1.2).

Therefore, the process if discretizing a color reduces to that of discretizing a function, then what is the most natural technique? A simple solution consists in using point sampling as we did in the discretization of the parabola in example 2. In this case another question arises: How many samples should we take?

Understanding this problem in its generality and devising adequate and robust solutions is a difficult task. In fact this is a topic to be covered in this book. In the case of color, it is possible to devise a simple solution going back to the physical universe. How does the human eye processes color? From the theory of Young-Helmholtz the eye discretizes color using three samples of the spectral distribution function: One sample of low frequency (Red), one sample of medium frequency (Green), and another sample of high frequency (Blue).

This means that from the perceptual point of view it is enough to use three samples of the spectral distribution function in order to represent color. This means that the representation of the color space (which is an infinite dimensional function space) is given by the Euclidean space $\mathbb{R}^3$. Note that this representation is completely justified from the point of view of color perception on the physical universe. It is far away from being a good mathematical representation.

Exact reconstruction of the original spectral color from the three RGB samples is in general an impossible task. Nevertheless, the color sampling problem has been posed in a way that we do not need to guarantee an exact reconstruction: We should be able to provide a reconstruction process in which the reconstructed color is perceptually equal to the color defined by the original spectral distribution function. This problem is called *perceptual color reconstruction*. Perceptual reconstruction of color is the technology behind many color reconstruction devices nowadays, such as color television: The TV set reconstructs colors which are only perceptually equal to the original colors captured from the video camera.

1.3 Functions and Computational Mathematics

Functions play a distinguished role in computational mathematics. In fact they constitute the basic tools of mathematical modeling: The relation between magnitudes of the physical world is, in most cases, described by functions. Notice that the two examples of rigid body motion and color reconstruction were reduced to the problem of representation and reconstruction of functions.

Therefore, a highly relevant problem in computational mathematics turns out to be the problem of function representation and reconstruction.

1.3.1 Representation and Reconstruction of Functions

The simplest method to represent a function is to use point sampling. In the one-dimensional case a real function of one-variable $f: I \subset \mathbb{R} \to \mathbb{R}$ is discretized by taking a partition $t_1 < t_2 < \cdots < t_n$ of the domain interval I. The representation is given by the vector

$$f_n = (f(t_1), f(t_2,), \ldots, f(t_n)) \in \mathbb{R}^n .$$

In this way, the space of real functions defined on the interval I is represented by the Euclidean space $\mathbb{R}^n$.

Is this a good representation? This question is directly related with the problem of information loss in the representation of a function. A mathematical formulation of the above question would be: Is point sampling an exact representation? In other words, is it possible to reconstruct a function f from its representation vector? Note that in the case of point sampling a reconstruction technique consists of interpolating the points $(t_i, f(t_i))$, $i = 1, \ldots, n$. In general, this interpolation process gives only an approximation of the original function.

A final word about function reconstruction is in order. Inside the computer any function is in fact necessarily discretized. When we talk about a continuous function it means that we are able to evaluate the function at any point of its domain. That is, we have a representation of the function on the computer, along with a reconstruction technique that enables us to evaluate the function at any point.

1.3.2 *Specification of Functions*

In general when a function is specified on the computer the specification technique results in a representation of the function. The first step towards doing computation with the function consists in reconstructing it from its representation.

Therefore representation and reconstruction techniques are related with user interface issues in a computational system.

1.4 What is the Relation with Graphics?

A broad view of computer graphics would lead us to say that is the area that manipulates graphical objects. That is, in computer graphics we are interested in creating, processing, analyzing, and visualizing graphical objects.

In order to fully understand this description we must give a precise meaning to the concept of a graphical object. A simple and effective definition was given in [25]: A graphical object consists of a shape and its attributes. The shape defines the geometry and topology of the object, and the attributes describe the different properties associated with it, such as color, material, velocity, density, etc.

Mathematically, the shape is characterized as a subset $U \subset \mathbb{R}^n$, and the attributes are encapsulated into a vector valued function

$$f : U \subset \mathbb{R}^n \to \mathbb{R}^p .$$

The dimension of the set U is the dimension of the graphical object, and $\mathbb{R}^n$ is the embedding space. The object is denoted by $\mathcal{O} = (U, f)$.

This definition encompasses different signals such as video, audio and image, and also curves, surfaces, and volumetric models. According to the abstraction levels we have described the main problems of computer graphics such as description, representation, reconstruction, and implementation of graphical objects. We should remark that visualization of graphical objects, an important part of the computer graphics applications, is in fact a reconstruction problem.

Example 4 (Image and Digital Image). The physical model of an image is well represented by a photography. Analyzing a photography we observe a geometric support (the paper where it is printed), and associated with each point of this support we have a color information.

Therefore, from the mathematical point of view we can model an image as a function $f: U \subset \mathbb{R}^2 \to \mathscr{C}$, where $\mathscr{C}$ is a representation of the color space. Using the RGB color representation of example 3 we have $\mathscr{C} = \mathbb{R}^3$.

We can easily obtain a representation of an image using point sampling. We take partitions $\Delta X = \{i\Delta x\, ;\ i \in \mathbb{Z},\ \Delta x > 0\}$ and $\Delta Y = \{j\Delta y\, ;\ j \in \mathbb{Z},\ \Delta y > 0\}$ of the x and y axes of the Euclidean plane, and we define a two-dimensional lattice of the plane by taking the cartesian product $\Delta X \times \Delta Y$. This lattice is characterized by the vertices $(i\Delta x, j\Delta y)$ which are denoted simply by (i,j). The image f is represented by the matrix

$$f(i,j) = f(i\Delta x, j\Delta y)\ .$$

The values of Δx and Δy represent the resolution of the representation. An important problem consists in devising the best resolution that is appropriate for each application.

1.4.1 Description of Graphical Objects

By description of a graphical object we mean the definition of its shape and attributes on the mathematical universe.

There are basically three different methods to describe the shape of a graphical object.

- Parametric description;
- Implicit description;
- Algorithmic description.

In the parametric description the object shape is defined as the image of a function $f: U \subset \mathbb{R}^m \to \mathbb{R}^n$. If U is an m-dimensional subset of $\mathbb{R}^m$, the graphical object has dimension m.

In the implicit description the object shape $U \subset \mathbb{R}^m$ is described as the zero set of a function $F: \mathbb{R}^m \to \mathbb{R}^k$. More precisely,

$$U = \{x \in \mathbb{R}^m\ ;\ F(x) = 0\}\ .$$

This set is denoted by $F^{-1}(0)$, and it is called the inverse image of 0 by the function f.

In the algorithmic description the shape is described by a function (either implicitly or parametrically) which is defined using an algorithm in some virtual machine (see [5]). Examples of functions defined algorithmically are found in the description of fractals. Other common example in computational mathematics are the objects defined using differential equations, where we must use an algorithm to solve the equation.

An important remark is in order. The attributes of a graphical object are described by a function. Also as discussed above, in general, the shape of the object is described by functions. Therefore, both the shape and attributes of a graphical object are described by functions. The study of graphical objects reduces to the problem of function description, representation, and reconstruction.

1.5 Where do Wavelets Fit?

In order to understand the role of the wavelets in the scenario of computational mathematics, even without understanding what a wavelet is, we must remember that our major concern is the description, representation, and reconstruction of functions.

The different uses of wavelets in computational mathematics, and in particular in computer graphics, are related with two facts:

- Representation and reconstruction of functions, a problem that has been posed above;
- Multiresolution representation, a problem that consists in representing the graphics object in different resolutions.

1.5.1 Function Representation Using Wavelets

An efficient process to represent a function consists of decomposing it into simpler functions. In other words, we must devise a collection $\{g_\lambda\}$ of functions $\lambda \in \Omega$, where the parameter space Ω is discrete, in such a way that every function f can be expressed by a sum

$$f = \sum_{\lambda} c_\lambda g_\lambda \,. \tag{1.1}$$

In this case, the coefficient sequence (c_λ) constitutes the representation of the function f, $f \mapsto (f_\lambda)_{\lambda \in \Omega}$. The function f is reconstructed from its representation coefficients using Eq. (1.1). Wavelets constitute a powerful tool to construct families g_λ, and compute the representation coefficients c_λ.

1.5.2 Multiresolution Representation

Our perception of the physical world occurs in different scales. In order to recognize a house it is not necessary to have details about its roof, doors, or windows. The identification is done using a macroscopic scale of the house. Different information about the house can be attained by using finer scales, where the measurements of the details have the same order of magnitude of the scale used.

This multiple scale approach is the desired one when we study functions. A good representation of a function should enable us to devise representations in different scales. Wavelets turn out to be a very effective tool in order to achieve multiresolution representations.

1.6 About these Book

This book describes a guided journey to the wonderland of the wavelets. The starting point of our journey is the kingdom of Fourier analysis. The book has an introductory flavor. We have tried to use mathematics as a language which has the adequate semantics to describe the wavescape along the trip. Therefore, in order to make the text easier to read we decided to relax with the mathematical syntax. To cite an example, it is very common to refer to some "space of functions" without making the concept precise. Inner products and norms on these spaces will be used. Also, we will refer to linear operators and other operations on function spaces relaxing about several hypothesis that would be necessary to confirm the validation of some assertions.

There are several books that cover these syntax issues in detail in the literature. We have tried to write a logfile of our journey which contains somehow the details without going deep into the mathematical rigor. We are more interested into the semantics of the different facets of wavelet theory, emphasizing the intuition over the mathematical rigor.

In spite of the huge amount of material about wavelets in the literature, this book covers the subject with a certain degree of originality on what concerns the organization of the topics. From the point of view of the abstraction paradigm of the four universes, we will cover the role of wavelets on function representation and reconstruction. We will also discuss some implementation issues along the way.

This book originated from our need to teach wavelets to students originated from both mathematical and computer science courses.

1.7 Comments and References

The abstraction levels of the four universes described in this section are based on [46]. A generalization of them along the lines presented in this section was published in [26].

Many good books on wavelets have been published. They emphasize different aspects of the theory and applications. We will certainly use material from these sources, and we will cite them along our journey.

Chapter 2
Function Representation and Reconstruction

In the previous chapter we concluded that one of the major problems in computational mathematics is related to function representation and reconstruction. In this chapter we will give more details about these two problems in order to motivate the study of wavelets.

2.1 Representing Functions

Functions must be discretized so as to implement them on the computer. Also, when a function is specified in the computer the input is a representation of the function. As an example, numerical methods that solve differential equations (Runge-Kutta, finite differences, finite elements, etc.) compute in fact a representation of the solution.

Associated with each representation technique we must have a reconstruction method. These two operations enable us to move functions between the mathematical and the representation universes when necessary. As we will see in this chapter, the reconstruction methods are directly related with the theory of function interpolation and approximation.

J. Gomes, L. Velho, *From Fourier Analysis to Wavelets*,
IMPA Monographs 3, DOI 10.1007/978-3-319-22075-8_2

2.1.1 The Representation Operator

We denote by $\mathcal{S}$ a space of sequences $(c_j)_{j\in\mathbb{Z}}$ of numbers[1]. We admit that $\mathcal{S}$ has a defined norm which allows us to compute the distance between two sequences of the space. If $c = (c_j)j \in \mathbb{Z}$, its norm is indicated by $||c|| = ||(c_j)j \in \mathbb{Z}||$.

A *representation* of a space of functions $\mathcal{F}$ is an operator $R: \mathcal{F} \to \mathcal{S}$ into some space of sequences. For a given function $f \in \mathcal{F}$, its representation $R(f)$ is a sequence

$$R(f) = (f_j)_{j\in\mathbb{Z}} \in \mathcal{S} .$$

R is called the *representation operator*. We also suppose that $\mathcal{F}$ has a norm, $||\ ||$, and that R preserves norms, that is

$$||R(f)|| = ||f|| ,$$

or that R satisfies some stability condition, such as

$$||R(f)|| \leq C||f|| .$$

When R is linear and continuous, we have a linear representation.

The most important examples of representations occur when the space of functions $\mathcal{F}$ is a subspace of the space $\mathbf{L}^2(\mathbb{R})$, of square integrable functions (finite energy),

$$\mathbf{L}^2(\mathbb{R}) = \left\{ f: \mathbb{R} \to \mathbb{R} ;\ \int_{\mathbb{R}} |f(t)|^2 dt < \infty \right\} ,$$

and the representation space $\mathcal{S}$ is the space ℓ^2 of the square summable sequences,

$$\ell^2 = \left\{ (x_i)_{i\in\mathbb{Z}},\ ;\ x_i \in \mathbb{C}, \text{and} \sum_{i=-\infty}^{+\infty} |x_i|^2 < \infty \right\} .$$

When the representation operator is invertible, we can reconstruct f from its representation sequence: $f = R^{-1}((f_i)_{i\in\mathbb{Z}})$. In this case, we have an *exact representation*, also called *ideal representation*. A method to compute the inverse operator gives us the reconstruction equation. We should remark that in general invertibility is a very strong requirement for a representation operator. In fact weaker conditions such as invertibility on the left suffice to obtain exact representations.

In case the representation is not exact we should look for other techniques to compute approximate reconstruction of the original function. There are several representation/reconstruction methods in the literature. We will review some of these methods in this chapter.

---

[1]By numbers we mean a real or complex number.

2.2 Basis Representation

A natural technique to obtain a representation of a space of functions consists in constructing a basis of the space. A set $B = \{e_j; j \in \mathbb{Z}\}$ is a basis of a function space $\mathcal{F}$ if the vectors e_j are linearly independent, and for each $f \in \mathcal{F}$ there exists a sequence $(\alpha_j)_{j\in\mathbb{Z}}$ of numbers such that

$$f = \sum_{j\in\mathbb{Z}} \alpha_j e_j \ . \tag{2.1}$$

The above equality means the convergence of partial sums of the series in the norm of the space $\mathcal{F}$.

$$\lim_{n\to\infty} \left\| f - \sum_{j=-n}^{n} \alpha_j e_j \right\| = 0$$

We define the representation operator by

$$R(f) = (\alpha_j)_{j\in\mathbb{Z}} \ .$$

Equation (2.1) reconstructs the function f from the representation sequence (α_j).

We must impose additional hypothesis on the representation basis in order to guarantee unicity of the representation. A particular case of great importance (in fact the only one we will be using in this book) occurs when the space of functions has an inner product and it is complete in the norm induced by this inner product. These spaces, called Hilbert spaces, possess special basis as we will describe below.

2.2.1 Complete Orthonormal Representation

A collection of functions $\{\varphi_n; n \in \mathbb{Z}\}$ on a separable Hilbert space $\mathcal{H}$ is a *complete orthonormal set* if the conditions below are satisfied:

1. **Orthogonality:** $\langle \varphi_m, \varphi_n \rangle = 0$ if $n \neq m$;
2. **Normalization:** $||\varphi_n|| = 1$ for each $n \in \mathbb{Z}$;
3. **Completeness:** For all $f \in \mathcal{H}$, and any $\varepsilon > 0$, there exists $N > 0$, $N \in \mathbb{Z}$ such that

$$\left\| f - \sum_{k=-N}^{N} \langle f, \varphi_k \rangle \varphi_k \right\| < \varepsilon \ .$$

The third condition says that linear combinations of functions from the complete orthonormal set can be used to approximate arbitrary functions from the space. Complete orthonormal sets are also called *orthonormal basis* of Hilbert spaces.

An orthonormal basis $\{\varphi_j\}$ defines a representation operator $R: \mathcal{H} \to \ell^2$, $R(f) = (f_j) = (\langle f, \varphi_j \rangle)$, which is invertible. Therefore, the representation is exact. The reconstruction of the original signal is given by

$$f = \sum_{k=-\infty}^{+\infty} \langle f, \varphi_k \rangle \varphi_k .$$

It is easy to see that the orthogonality condition implies that the elements φ_n are linearly independent. This implies in particular that the representation sequence $(\langle f, \varphi_k \rangle)$ is uniquely determined by f. This representation preserves the norm. That is

$$||R(f)||^2 = \sum_{k \in \mathbb{Z}} \langle f, \varphi_k \rangle^2 = ||f||^2 . \tag{2.2}$$

This expression is called *Plancherel equation.*

2.3 Representation by Frames

The basic idea of representing functions on a basis consists in decomposing it using a countable set of simpler functions.

The existence of a complete orthonormal set and its construction is in general a very difficult task. On the other hand, orthonormal representations are too much restrictive and rigid. Therefore, it is important to obtain collections of functions $\{\varphi_n; n \in \mathbb{Z}\}$ which do not constitute necessarily an orthonormal set and are not even linearly independent, but can be used to define a representation operator. One such collection is constituted by the *frames.*

Consider a space $\mathcal{H}$ with an inner product $< >$. A collection of functions $\{\varphi_n; n \in \mathbb{Z}\}$ is a frame if there exist constants A and B satisfying $0 < A \leq B < +\infty$, such that for all $f \in \mathcal{H}$, we have

$$A||f||^2 \leq \sum_{n=-\infty}^{\infty} |\langle f, \varphi_n \rangle|^2 \leq B||f||^2 . \tag{2.3}$$

The constants A and B are called *frame bounds.* When $A = B$ we say that the frame is *tight.* From the Plancherel formula (2.2) it follows that every orthonormal set is a tight frame with $A = B = 1$. Nevertheless, there exist tight frames which are not orthonormal basis. The following statement is true:

Theorem 1. *If $\mathcal{B} = \{\varphi_n; n \in \mathbb{Z}\}$ is a tight frame with $A = 1$ and $||\varphi_n|| = 1, \forall n$, then $\mathcal{B}$ is an orthonormal basis.*

If a frame is tight, it follows from (2.3) that

$$\sum_{j\in\mathbb{Z}} |\langle f, \varphi_j\rangle|^2 = A||f||^2 .$$

Using the polarization identity, we obtain

$$A\langle f, g\rangle = \sum_{j\in\mathbb{Z}} \langle f, \varphi_j\rangle \langle \varphi_j, g\rangle ,$$

that is,

$$f = A^{-1} \sum_{j\in\mathbb{Z}} \langle f, \varphi_j\rangle \varphi_j .$$

The above expression (although deduced in the weak sense) shows how we can obtain approximations of a function f using frames. In fact, it motivates us to define a representation operator R analogous to what we did for orthonormal basis:

$$R(f) = (f_j)_{j\in\mathbb{Z}}, \quad \text{where} \quad f_j = \langle f, \varphi_j\rangle .$$

We should remark that this operator in general is not invertible. Nevertheless it is possible to reconstruct the signal from its representation $R(f)$, as we will show below.

From Eq. (2.3) it follows that the operator R is bounded: $||Rf||^2 \leq B||f||^2$. The adjoint operator R^* de R is easily computed:

$$\langle R^*u, f\rangle = \langle u, Ru\rangle = \sum_{j\in\mathbb{Z}} u_j \overline{\langle f, \varphi_j\rangle} = \sum_{j\in\mathbb{Z}} u_j \langle \varphi_j, f\rangle ,$$

therefore

$$R^*u = \sum_{j\in\mathbb{Z}} u_j \varphi_j .$$

(It can be shown the convergence is also true on norm.) By the definition of R we have

$$\sum_{j\in\mathbb{Z}} |\langle f, \varphi_j\rangle|^2 = ||Ff||^2 = \langle F^*Ff, f\rangle .$$

On the other hand, since $||F^*|| = ||F||$, we have $||F^*u|| \leq B^{1/2}||u||$. We conclude that

$$AI \leq F^*F \leq BI .$$

In particular, it follows that the operator $F^*F\colon \mathscr{F} \to \mathscr{F}$ is invertible.

The above results allow us to obtain an expression to reconstruct the function f from its representation $R(f)$. We will state the result without proof: Applying the operator $(F^*F)^{-1}$ to the elements φ_j of the frame, we obtain a family of functions $\tilde{\varphi}_j$,

$$\tilde{\varphi}_j = (F^*F)^{-1}\varphi_j \, .$$

This family constitutes a frame with bounds given by

$$B^{-1}||f||^2 \leq \sum_{j\in\mathbb{Z}} |\langle f, \tilde{\varphi}_j\rangle|^2 \leq A^{-1}||f||^2 \, .$$

Therefore we can associate with the frame $\{\tilde{\varphi}_j\}$ a representation operator $\tilde{R}\colon \mathscr{F} \to \ell^2(\mathbb{Z})$, defined in the usual way $(\tilde{R}f)_j = \langle f, \tilde{\varphi}_j\rangle$. The frame $\{\tilde{\varphi}_j\}$ is called *reciprocal frame* of the frame $\{(\varphi_j)_{j\in\mathbb{Z}}\}$. It is easy to prove that the reciprocal frame of $\{\tilde{\varphi}_j\}$ is the original frame $\{(\varphi_j)\}$.

The following identity can also be proved for all functions $f \in \mathscr{F}$:

$$f = \sum_{j\in\mathbb{Z}} \langle f, \varphi_j\rangle \tilde{\varphi}_j = \sum_{j\in\mathbb{Z}} \langle f, \tilde{\varphi}_j\rangle \varphi_j \, . \tag{2.4}$$

The first equality gives us a reconstruction equation of the function f from its representation sequence $R(f) = (\langle f, \varphi_j\rangle)_{l\in\mathbb{Z}}$. Note that the second equation in (2.4) gives a method to describe a function f as a superposition of the elements of the original frame. Because of this equation the reciprocal frame is also called *dual frame*.

2.4 Riesz Basis Representation

If a collection $\mathscr{B} = \{\varphi_n; n \in \mathbb{Z}\}$ is a frame, and the functions φ_j are linearly independent, we have a *Riesz basis*. Therefore if $\{e_n\}$ is a Riesz basis, for any $f \in \mathscr{H}$ we have

$$A||f||^2 \leq \sum_{n} |\langle f, e_n\rangle|^2 \leq B||f||^2 \, .$$

For each function $f \in \mathbf{L}^2(\mathbb{R})$, we define its representation

$$\sum_{k} \langle f, e_k\rangle e_k \, ,$$

on the Riesz basis $\{e_k\}$.

If L is a continuous linear operator with a continuous inverse, then L maps any orthonormal basis of a Hilbert space onto a Riesz basis. Moreover, any Riesz basis can be obtained using this method.

Also, it can be proved that from a Riesz basis $\{e_n\}$ of a Hilbert space $\mathcal{H}$, we can construct an orthonormal basis $\{\tilde{e}_n\}$ of $\mathcal{H}$. A proof of this fact can be found in [20], p. 139.

From the above remarks, we see that Riesz bases constitute the basis of a Hilbert space that is closest to complete orthonormal sets.

2.5 Representation by Projection

For a given closed subspace V of a Hilbert Space $\mathcal{F}$, the representation by projection consists in taking a representation of a function $f \in \mathcal{F}$ as its unique orthogonal projection onto V (see Fig. 2.1).

If $\{\varphi_k\}$ is an orthonormal basis of V, then a representation of f is given by

$$R(f) = \mathrm{Proj}_V(f) = \sum_k \langle f, \varphi_k \rangle \varphi_k .$$

This representation is not exact, unless we have $V = \mathcal{F}$. The best reconstruction we can have consists in finding the unique vector of V which is closest to f. This vector can be computed using optimization techniques.

An important particular case occurs when we have a family of subspaces $\{V_j\}$ such that $\overline{\bigcup V_j} = \mathcal{F}$, and the family of representations $\mathrm{Proj}_{V_j}(f)$ converges to the function f. An example occurs when each subspace V_j has finite dimension, as described in the next section.

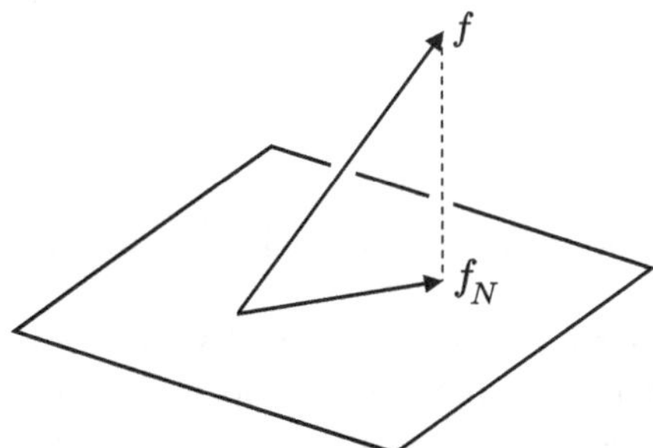

Fig. 2.1 Galerkin representation of a function f

2.6 Galerkin Representation

The Galerkin method computes a representation $R(f) = (f_j), j = 0, 1, \ldots, N-1$ in such a way that there exists a reconstruction equation

$$f_N(t) = \sum_{k=0}^{N-1} a_N(k)\phi_{N,k}(t) ,$$

which approximates the function f, in norm, when $N \to \infty$. That is,

$$||f - f_N|| \to 0, \quad \text{if} \quad N \to \infty .$$

The representation f_N is therefore computed in such a way to minimize the norm $||f - f_N||$. Geometrically, the functions

$$\phi_{N,0}, \phi_{N,1}, \ldots, \phi_{N,N-1}$$

generate a subspace V of the space of functions, and the representation f_N is the orthogonal projection of f onto V (see Fig. 2.1).

The coefficients $a_N(k)$ are computed using a dual set $\varphi_{N,k}$ of the set $\phi_{N,k}$:

$$\langle \phi_{N,k}, \varphi_{N,j} \rangle = \delta_{i,j}, \qquad a_N(k) = \langle f, \varphi_{N,k} \rangle .$$

The functions $\varphi_{N,k}$ are called *sampling functions*, *analysis functions*, or *representation functions*. The functions $\phi_{N,k}$ are called *synthesis functions* or *reconstruction functions*.

It is important to remark that the reconstruction functions are not uniquely determined. Also, changing these functions imply in a change of the reconstructed function. If $\phi_{N,k}$ is an orthonormal set, then the synthesis functions may be used as analysis functions, that is,

$$a_N(k) = \langle f, \phi_{N,k} \rangle .$$

The Galerkin representation is also known in the literature as *representation by finite projection*.

Galerkin representation is very important in devising numerical methods because they use finite dimensional representation spaces, which is very suitable for computer implementations. This method is used in different representations of functions by piecewise polynomial functions in the theory of approximation.

2.7 Reconstruction, Point Sampling and Interpolation

The well-known example of a Galerkin representation consists in taking a finite point sampling of a function $f: [a, b] \to \mathbb{R}$. We take a partition $a = t_0 < t_1, \cdots, < t_n = b$ and define

$$R(f) = (f(t_0), \ldots, f(t_n)) \in \mathbb{R}^n .$$

In this case, the Dirac "functions" $\delta(t - t_k)$, $k = 0, 1, \ldots, n$ constitute the sampling functions

$$f(t_k) = \langle f, \delta(t - t_k) \rangle = \int_{-\infty}^{+\infty} f(u)\delta(t - t_k)du .$$

We can choose several different reconstruction functions from the above representation. We will study some classical reconstruction basis below.

2.7.1 Piecewise Constant Reconstruction

In this reconstruction each synthesis function is given by the characteristic functions of the partition intervals.

$$\phi_{N,k} = \chi_{[t_k, t_{k+1}]} = \begin{cases} 0 & \text{if } x < t_k \text{ or } t > t_{k+1} \\ 1 & \text{if } x \in [t_k, t_{k+1}] \end{cases} \quad (2.5)$$

The graph of this function is shown in Fig. 2.2(a). Geometrically, the reconstructed function is an approximation of the function f by a function which is constant in each interval of the partition (see Fig. 2.2(b)).

In the literature of signal processing the synthesis function used here is called *box function*. This method reconstructs the functions approximating it by a discontinuous functions. The discontinuities in the reconstructed function introduce high frequencies in the reconstructed signal (see Chap. 7 of [24]).

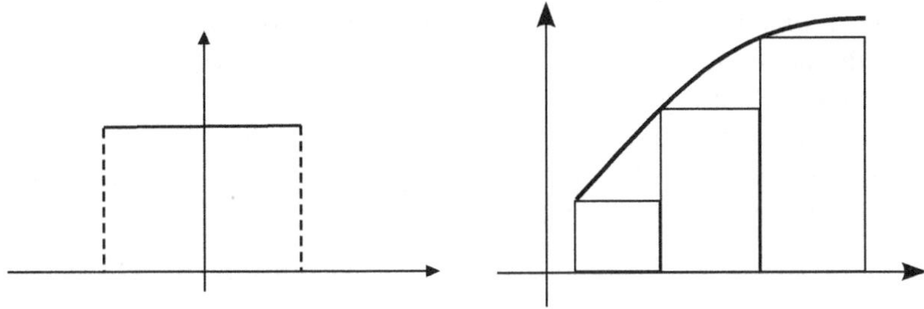

Fig. 2.2 **a** Graph of the synthesis function. **b** Reconstruction

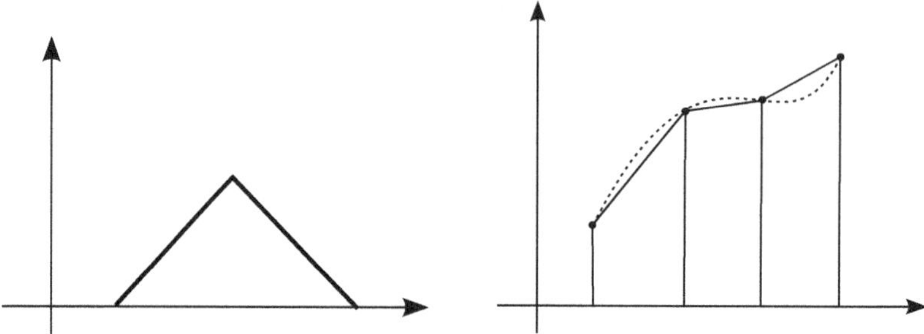

Fig. 2.3 **a** Graph of the synthesis function. **b** Reconstruction

2.7.2 *Piecewise Linear Reconstruction*

In this case the synthesis functions $\phi_{N,k}$ are given by

$$\phi_{N,k} = \begin{cases} 0 & \text{if } x < t_{k-1} \text{ or } t > t_{k+1} \\ \dfrac{t - t_{k-1}}{t_k - t_{k-1}} & \text{if } x \in [t_{k-1}, t_k] \\ \dfrac{t_{k+1} - t}{t_{k+1} - t_k} & \text{if } x \in [t_k, t_{k+1}] \end{cases} \tag{2.6}$$

The graph of this function is shown in Fig. 2.3(a). Geometrically, the reconstructed function is an approximation of the function f by a continuous function which is linear in each interval of the partition used in the sampling process. Fig. 2.3(b) shows, in dotted lines, the graph of the original function, and the graph of the reconstructed function using a continuous line.

Higher Order Reconstruction

We could continue with these reconstruction methods using polynomial functions of higher degree defined on each interval of the sampling partition. The differentiability class of these functions on the whole interval can be controlled according to the degree of the polynomials. The spaces of splines is the natural habitat for these reconstruction methods.

We should remark that in the Galerkin representation using point sampling the reconstruction methods consist in interpolating and approximating the original function from its samples. A natural question would be:

Question 2.1. Is it possible to obtain exact reconstruction using the Galerkin representation by point sampling?

Note that exact reconstruction in this representation has an interesting geometric meaning: *Find an interpolation method that is able to reconstruct the function f exactly from the set of its samples $f(t_i)$.* We will return to this question in the next chapter.

2.8 Multiresolution Representation

We perceive the world through a multiscale mechanism. First we use a coarse scale to recognize the object, then we use finer scales in order to discover its distinct properties in detail. As an example, the identification of a house can be done in a very coarse scale, but finer scales are necessary in order to observe details about the windows, doors, floor, and so forth.

Therefore, it is natural that we look for multiscale representation of functions. That is, we are interested in obtaining a family of representations that could represent the function at distinct scales. At the same time we need techniques that allow us to change between representations on different scales.

This can be achieved by using nested representation spaces. Consider a sequence of closed subspaces $\{V_j\}_{j\in\mathbb{Z}}$ such that

$$\cdots V_{j+1} \subset V_j \subset V_{j-1} \cdots , \quad \forall \in \mathbb{Z} .$$

And a family of representation operators $P_j \colon V \to V_j$ such that

$$||v - P_j(v)|| \leq C \inf_{u\in V_j} ||v - u|| , \tag{2.7}$$

where C does not depend on j. The proximity condition of Eq. (2.7) is trivially satisfied if V_j is a closed subspace of a Hilbert space, and P_j is the orthogonal projection onto V_j, as illustrated in Fig. 2.4.

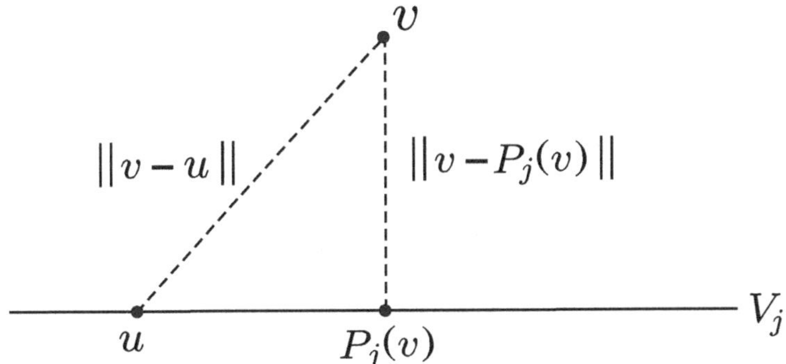

Fig. 2.4 Representation operator

Note that from Eq. (2.7) we have

$$||P_j(v)|| \leq c||v|| \tag{2.8}$$

$$P_j(v) = v, \quad \forall v \in V_j \text{ (i.e. } P_j^2 = P_j) . \tag{2.9}$$

Also we require a natural commutativity condition $P_j \circ P_{j-1} = P_j$, which guarantees that the different representations match each other.

Intuitively each representation space V_{j-1} contains more details (finer scale) than the space V_j. This can be stated precisely by introducing the operator Q_j: $V \to V_{j-1}$, defined by

$$Q_j(v) = P_{j+1}(v) - P_j(v) .$$

If $W_j = \text{Image}(Q_j) = Q_j(V_j)$, it follows that

$$V_{j-1} = V_j + W_j . \tag{2.10}$$

That is, $Q_j(v)$ is the detail we must add to the representation space V_j to obtain V_{j+1}. For this reason, Q_j is called a *refinement operator*.

Iterating Eq. (2.10) we have the decomposition equation

$$V_J = V_{j_0} + \left(W_{j_0} + \cdots + W_{J+1}\right) , \tag{2.11}$$

which says that a finer representation can be obtained from a previous one, by adding details. Equation (2.11) can be rewritten using operators:

$$P_j(v) = P_{j_0}(v) + \sum_{j=J+1}^{j_0} Q_j(v) . \tag{2.12}$$

In order to be able to decompose any element v of V we impose some additional hypothesis on the representation operators P_j:

1. $P_j(v) \to v$ if $j \to -\infty$;
2. $P_j(v) \to 0$ if $j \to +\infty$.

From the proximity condition of Eq. (2.7), these two conditions are equivalent to

1. $\bigcup_{j\in\mathbb{Z}} V_j$ is dense on V;
2. $\bigcap_{j\in\mathbb{Z}} = \{0\}$.

By taking the limit, $j_0 \to +\infty$ and $j \to -\infty$, in (2.12) we have

$$v = \sum_{j\in\mathbb{Z}} Q_j(v), \quad \forall v \in V .$$

That is, any vector v is the sum of all of its details in the different representations.

In the particular case where the representation operators P_j are orthogonal projections over closed Hilbert spaces, the sum in Eq. (2.11) is in fact a direct sum, and the spaces V_j and W_j are orthogonal.

Techniques to compute nested sequences of multiresolution representation are closely related with wavelets.

2.9 Representation by Dictionaries

The problem of obtaining a representation of a function can be well understood using the metaphor of natural language. When we write, we have an idea and we must represent it using words of the language. If we have a rich vocabulary, we will have a great conciseness power to write the idea, on the contrary, we will have to use more words to express the same idea. Based on this metaphor, S. Mallat [38] introduced in the literature a representation method based on dictionary of functions.

A dictionary in a function space H is a family $\mathscr{D} = (g_\lambda)_{\lambda\in\Gamma}$ of vectors in H such that $||g_\lambda|| = 1$. This family is not necessarily countable. A representation of a function in a dictionary $\mathscr{D}$ is a decomposition

$$f = \sum_{\lambda\in\Gamma} \alpha_\gamma g_\gamma ,$$

such that $(\alpha_\gamma) \in \ell^2$. The rationale behind the representation method consists in constructing extensive dictionaries and devising optimal representation techniques that allow us to represent a function using a minimum of words from the dictionary. Therefore, Representation using dictionaries allows us the use of a great heterogeneity in the reconstruction functions, which makes the representation/reconstruction process very flexible.

Note that distinct functions use different dictionary vectors in their representation which makes the representation process nonlinear. In [37] several dictionary systems are described, as well as techniques to compute dictionary based representations. One basic strategy to compute a representation for a function f is described below.

Let f_M be the projection of f over the space generated by M vectors from the dictionary, with index set I_M:

$$f_M = \sum_{m\in I_M} \langle f, g_m\rangle g_m .$$

The error e_M in the approximation is the sum of the remaining coefficients

$$e_M = ||f - f_M||^2 = \sum_{m\notin I_M} |\langle f, g_m\rangle|^2 .$$

To minimize this error the indices in I_M must correspond to the M vectors that have the largest inner product amplitude $|\langle f, g_m\rangle|$. These are the vectors that have a better correlation with f, that is, the vectors that best match the features of f. Certainly the error is smaller than the error resulting from a linear approximation where the decomposition vectors of the representation do not vary with f.

The above ideas lead to an algorithm to compute the representation. For details see [38].

2.10 Redundancy in the Representation

The representation of a function is not unique in general. Besides non-unicity, we can have a redundancy for a given representation. This occurs, for instance, in the representation using frames. In fact, if the frame $(\varphi_j)_{j\in\mathbb{Z}}$ is an orthonormal basis, the representation operator

$$R\colon \mathscr{H} \to \ell^2, \qquad (Rf)_j = \langle f, \varphi_j\rangle$$

is an isometry, and the image of $\mathscr{H}$ is the whole space ℓ^2. The reconstruction equation

$$f = \sum_{j\in\mathbb{Z}} \langle f, \varphi_j\rangle \tilde{\varphi}_j$$

computes each vector in such a way that there is no correlation between the coefficients.

If the frame is not constituted by linearly independent vectors, there exists a redundancy in the representation, which corresponds to a certain correlation between the elements of the representation sequence. This redundancy can be used to obtain a certain robustness in the representation/reconstruction process. An interesting discussion of the redundancy in frame representation can be found in [20], p. 97.

2.11 Wavelets and Function Representation

We can summarize what we have done up to this point in the statement that our main concern is the study of representation and reconstruction of functions. Therefore our slogan at this point should be "all the way to function representation and reconstruction techniques."

In fact, we need to develop techniques that allow us to construct representation tools (basis, dictionaries, frames, multiscale representations, etc.) which are flexible, concise, and allow for robust reconstruction.

This leads us naturally to the study of function variation as a strategy to detect the different features of a given function, and take them into consideration when representing the function. We will start our journey on this road from the next chapter on. It will take us from the kingdom of Fourier to the wonderland of wavelets.

2.12 Comments and References

Multiscale representation was introduced by S. Mallat in [36] in the context of the Hilbert spaces $L^2(\mathbb{R})$ of functions, and orthonormal projection. The more general introduction we gave in Sect. 2.8 of this chapter was based on [10]. A detailed description of Mallat's work will be given later on.

A detailed study of representation and reconstruction using frames is found in [20], pp. 57–63, and also in Chap. V of [37].

Chapter 3
The Fourier Transform

In order to devise good representation techniques we must develop tools that enable us to locate distinguished features of a function. The most traditional of these tools is the Fourier Transform which we will study in this chapter. The study of Fourier transform, its strength and limitations, is the starting point of our journey to the wavelets.

3.1 Analyzing Functions

To detect features of a function we must analyze it. This occurs in our everyday routine: signals are analyzed and interpreted by our senses, a signal representation is grabbed from this analysis and it is sent to our brain. This is the process used in our perception of colors and sound.

Music, images, and other elements we interact with in our everyday life are characterized by functions: To each point on the space, and to each instant of time the function produces a certain output which we are able to detect. These functions are usually called *signals*.

The best way to analyze the features of a signal is by studying its frequencies. In an audio signal, for example, the frequencies are responsible for what we are accustomed to identify as an acute or grave sound. Also, the distinction from red to green is captured in the frequency of the associated electromagnetic wave.

3.1.1 Fourier Series

In order to analyze the frequency content of a function we must initially answer the following question: *What is the frequency of a function?* This is an easy task

J. Gomes, L. Velho, *From Fourier Analysis to Wavelets*,
IMPA Monographs 3, DOI 10.1007/978-3-319-22075-8_3

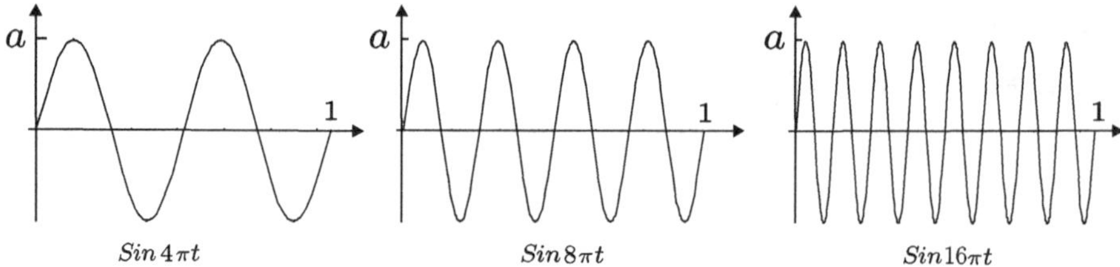

Fig. 3.1 Sine function with frequencies 2, 4, and 8

when the function is periodic. In fact given the function $f(t) = A\sin(2\pi\omega t)$, $A > 0$, the parameter A measures the amplitude (maximum and minimum values assumed by f), the parameter ω indicates how many complete cycles of period exist on the interval $[0, 1]$. This number is directly connected with the number of oscillations of the function in the unit of time, which is called the *frequency* of the function. Figure 3.1 shows the graph of f for different values of ω.

Consider a periodic function with period $L > 0$, that is, $f(t+L) = f(t)$. We denote by $L^2_T(\mathbb{R})$ the space of periodic functions of period T which are square integrable. That is

$$\int_{t_0}^{t_0+T} |f(t)|^2 dt < \infty\,.$$

The theory of Fourier series says that f can be decomposed as

$$f(s) = \sum_{j=-\infty}^{+\infty} a_j e^{i2\pi\omega_j s}, \quad a_j \in \mathbb{R}, \tag{3.1}$$

where $\omega_j = j/T$ is a constant. This decomposition of a periodic function f is called the *Fourier series* of f. It is well known that the family $\{e^{i2\pi\omega_j s},\ j \in \mathbb{Z}\}$ is a complete orthonormal set of the space $\mathbf{L}^2_T(\mathbb{R})$. Therefore Eq. (3.1) is an orthogonal basis representation of the function f.

In conclusion, the Fourier series shows that any periodic function can be decomposed as an infinite sum of periodic functions (sines and cosines). This decomposition makes it easy an analysis of the frequencies present on the function f: There exists a fundamental frequency ω, and all of the other frequencies are integer multiples ωj, $j \in \mathbb{Z}$, of this fundamental frequency.

The coefficient a_j in the Eq. (3.1) of the Fourier series measures the amplitude of the frequency component ω_j on the function f. In particular, if $a_j = 0$, this frequency is not present in the function. This frequency amplitude a_j is computed using the equation

$$a_j = \int_0^L f(u) e^{i2\pi\omega_j u} du\,, \tag{3.2}$$

where L is the period of the function. Note that Eq. (3.1) completely characterizes the function f by its frequencies. In other words, we have an exact representation of f.

3.1.2 Fourier Transform

The above portrait of the function f worked perfectly well: We were able to obtain an exact representation of the function f and this representation completely characterizes f by its frequencies. The only drawback is the fact that f was a periodic function.

Is it possible to extend the above results for non-periodic functions? In this case we do not have a discrete spectrum of well-defined frequencies as in Eq. (3.1). In fact, every function f defined by Eq. (3.1) is necessarily periodic. Nevertheless, we can use the Fourier series representation as an inspiration to introduce the concept of frequency for arbitrary functions.

Take $s = \omega_j$ in Eq. (3.2) which computes the amplitude of the frequency, and assume that the variable s takes any value. We obtain

$$a(s) = \int_{-\infty}^{+\infty} f(u)e^{i2\pi su}du\ . \tag{3.3}$$

Notice that we have changed the notation from a_j to $a(s)$.

The operation $f(u)e^{i2\pi su}$ in the integrand above is called *modulation* of the function f. The exponential is called the *modulating function.* For each s, $e^{i2\pi su}$ is a periodic function of frequency s, $s \in \mathbb{R}$. Therefore, for each $s \in \mathbb{R}$ Eq. (3.3) can be interpreted as an infinite weighted average of f using the modulating function as the weighting function.

The rationale behind the frequency computation in the modulation process can be explained as follows: When f has oscillations of frequencies s, or close to s, these frequencies result to be in resonance with the frequency s of the modulating function, therefore $a(s)$ assumes non-zero values. On the contrary, when the oscillations of f and the frequencies of the modulating function are completely distinct we have a cancellation effect and the integral of (3.3) is zero or close to zero. We conclude that $a(s)$ measures the occurrence of the frequency s on the function f. Since s varies in a continuum of numbers, it is interesting to interpret (3.3) as describing a *frequency density* of the function f. When $a(s) \neq 0$ this means that frequencies s occurs on f. The value of $a(s)$ is a measure of the occurrence of the frequency s on f.

We will change notation and denote $a(s)$ by $\hat{f}(s)$ or by $F(f)(s)$. Note that $\hat{f}$ is in fact an operator that associates with the function f the function $\hat{f} = F(f)$, defined by (3.3). Therefore we have an operator defined between two function spaces. What function spaces are these? It is possible to show that the operator is well defined for any function f satisfying the equation

$$||f|| = \int_{\mathbb{R}} |f(u)|^2 du < \infty . \tag{3.4}$$

These functions are called *square integrable functions* or functions with finite energy. This space is denoted by $\mathbf{L}^2(\mathbb{R})$. Equation (3.4) defines a norm on $\mathbf{L}^2(\mathbb{R})$, and it is easy to see that this norm is induced by the inner product.

$$\langle f, g \rangle = \int_{\mathbb{R}} f(u)\overline{g}(u)du ,$$

where $\overline{g}$ means the complex conjugate. It is well known that the space $\mathbf{L}^2(\mathbb{R})$ with this inner product is a Hilbert space, called the space of *square integrable functions*.

In sum, we have introduced an operator $F = \hat{f}: \mathbf{L}^2(\mathbb{R}) \to \mathbf{L}^2(\mathbb{R})$, defined by

$$F(f)(s) = \hat{f}(s) = \int_{-\infty}^{+\infty} f(u)e^{-i2\pi su}du . \tag{3.5}$$

This operator is called *Fourier Transform*. Note that we changed the signal of the exponent when passing from (3.3) to (3.5) so as to keep the definition compatible with the definition of the Fourier transform used in the literature.

Without going into too much details, we would like to clarify some mathematical points. The Fourier transform is well defined when the function f belongs to the space $L^1(\mathbb{R})$ of integrable functions. In this case it is easy to see that $\hat{f}$ is bounded and continuous. In fact,

$$|\hat{f}(w) - \hat{f}(s)| \leq \int_{\mathbb{R}} |f(t)|e^{i2\pi wt} - e^{i2\pi st}| \leq \left(\int_{\mathbb{R}} |f(t)dt\right) |w - s| .$$

The extension of f to the space $L^2(\mathbb{R})$ of square integrable functions is achieved by a limiting process: Any function $f \in L^2(\mathbb{R})$ can be approximated by integrable functions of finite energy. More precisely, for any $f \in \mathbf{L}^2(\mathbb{R})$, there exists a sequence $f_n \in L^1(\mathbb{R}) \cap L^2(\mathbb{R})$ such that

$$\lim_{n\to\infty} ||f_n - f|| = 0 .$$

We define $\hat{f}$ as the limit of the sequence $\hat{f}_n$. Why is it important to define the Fourier transform on $\mathbf{L}^2(\mathbb{R})$, instead of $L^1(\mathbb{R})$? We can give two main reasons for that:

- The Fourier transform of an integrable function is not necessarily integrable. Therefore F is not an operator $F: L^1(\mathbb{R}) \to L^1(\mathbb{R})$.
- The space $\mathbf{L}^2(\mathbb{R})$ has a richer structure than the space $L^1(\mathbb{R})$ because of its inner product. In fact it is a Hilbert space.

From the linearity of the integral it follows easily that the Fourier transform $F: L^2(\mathbb{R}) \to L^2(\mathbb{R})$ is linear. A fascinating result is that F is an isometry of $\mathbf{L}^2$. That is,

$$\langle f, g \rangle = \langle \hat{f}, \hat{g} \rangle, \quad \forall f, g \in \mathbf{L}^2(\mathbb{R}) . \tag{3.6}$$

This equation is known as *Parseval identity*. It follows easily from it that

$$||f||^2 = ||\hat{f}||^2 . \tag{3.7}$$

This equation is known as *Plancherel equation.*

The inverse of the Fourier transform is given by

$$F^{-1}(g)(t) = \int_{-\infty}^{+\infty} g(s)e^{i2\pi\omega t} ds . \tag{3.8}$$

That is, $F^{-1}(\hat{f}) = f$. Therefore, we have

$$f(t) = F^{-1}(\hat{f}) = \int_{-\infty}^{+\infty} \hat{f}(s)e^{i2\pi st} ds . \tag{3.9}$$

An important fact to emphasize is that we have two different ways to "read" Eq. (3.9) of the inverse transform:

1. It furnishes a method to obtain the function f from its Fourier Transform $\hat{f}$.
2. It allows us to reconstruct f as a superposition of periodic functions, $e^{i2\pi st}$, and the coefficients of each function in the superposition are given by the Fourier transform.

The second interpretation above shows that the equation of the inverse Fourier transform is a non-discrete analogous of Eq. (3.1) of the Fourier series.

3.1.3 Spatial and Frequency Domain

Analyzing from the mathematical point of view, the Fourier transform is an invertible linear operator on $\mathbf{L}^2(\mathbb{R})$. Nevertheless, in the applications we have a very interesting interpretation of it. A function $f: \mathbb{R} \to \mathbb{R}$ can be interpreted as associating with each value of t in the spatial domain $\mathbb{R}$ some physical magnitude $f(t)$. When we compute the Fourier transform of f, we obtain another function $\hat{f}(s)$ defined on $\mathbf{L}^2(\mathbb{R})$. In this case, for each value of the parameter $s \in \mathbb{R}$ the value $\hat{f}(s)$ represents the frequency density s in f. We interpret this by saying that $\hat{f}$ is a representation of the function f in the frequency domain. In summary, the Fourier transform changes a function from the spatial to the frequency domain.

Since in our case the spatial domain has dimension 1 it is common to interpret the variable t as time and call the spatial domain by the name of *time domain*.

Note that describing a function on the frequency domain allows us to obtain the frequency contents of the function. The frequency contents are closely related with the features carried by the function. As we have pointed out in the previous chapter, these features are important elements to obtain good function representation.

3.2 A Pause to Think

Our purpose in this section is to have a better understanding of the potentialities of the Fourier transform from the point of view of function representation and reconstruction. Equation (3.2) provides us with information about the frequencies of a periodic function. For this reason, it is called the *analysis equation*. Equation (3.1) allows us to obtain f from the coefficients a_j computed using the analysis equation. For this reason it is called a *synthesis equation*. From the point of view of the previous chapter the analysis equation computes a representation of the periodic function f by the sequence (a_j), $j \in \mathbb{Z}$. The synthesis equation provides us with a reconstruction technique. Note that in this case the reconstruction is exact.

When the function is not periodic we know that Eq. (3.5), which defines the Fourier transform, gives us an analysis of the frequencies of the function f. Equation (3.9) writes the function f as a superposition of a "continuum" of periodic functions. This equation plays the role of the reconstruction Eq. (3.1) of the function f (the analysis equation). Note that the analysis and synthesis equation associated with the Fourier transform are not discrete as in the case of the Fourier series, for periodic functions. Therefore we do not have a tool that allows us to represent and reconstruct arbitrary functions as in the case of the Fourier series.

One important question can be posed now: *How effective is the Fourier transform analysis of a function?* A discussion of this question will be done in the next section.

3.3 Frequency Analysis

In this section we study some examples that will give a better insight into the analysis provided by the Fourier transform.

Suppose that $\hat{f}(s_0) \neq 0$. From this we conclude that f has frequencies s_0 or close to s_0. In this case, the next step of analyzing f consists in determining the localization of these frequencies on the spatial domain of f. This localization is of great importance in several applications, and in particular in the problem of function representation. We can make an analogy using a radar metaphor: the existence of a certain frequency detects the presence of some object, and the localization of a frequency allows us to determine the object position.

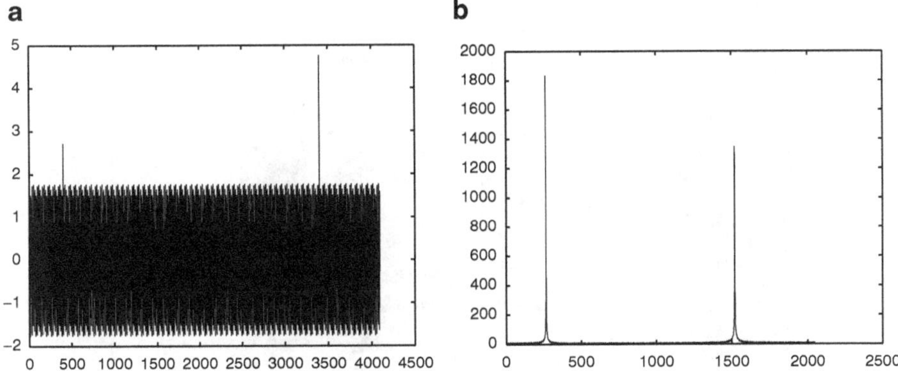

Fig. 3.2 A signal (a) and its Fourier transform (b)

In general localizing frequencies on the spatial domain using Fourier transform is impossible. This happens because the modulating function (exponential) used to measure the frequency density on the Fourier transform does not have compact support: the integral that defines the transform extends to the whole line. Therefore the only information carried from the fact that $\hat{f}(s) \neq 0$ is that the frequency s or frequencies close to s are present on the function f.

Example 5 (Signal with impulses). Consider the signal defined by the function

$$f(t) = \sin(2\pi 516.12t) + \sin(2\pi 2967.74t) + \delta(t - 0.05) + \delta(t - 0.42) .$$

It consists of a sum of two sines with frequencies 516.12 Hz and 2967.74 Hz, where we added two impulses of order 3 at the two different positions $t_0 = 0.05$ s and $t_1 = 0.42$ s. The graph of this signal is shown in Fig. 3.2(a). The graph of its Fourier transform is shown in Fig. 3.2(b).

Note that the Fourier transform detects well the two sine signals (the two spikes in Figure (b)). How about the impulses? It is easy to see from the definition of the Fourier transform, that translation of an impulse on the spatial domain introduces modulation by an exponential on the frequency domain. Therefore the modulation introduced by the two impulses generate a superposition of frequencies. These frequencies do not appear in Fig. 3.2(b) because of a scale problem. Figure 3.3 plots the graph in a different scale so as to make it possible to observe the superposed frequencies corresponding to the two impulses.

In conclusion the Fourier transform does not provide clear information about the two impulses.

The difficulty in localizing frequencies on the spatial domain is one of the major weaknesses of the Fourier transform in analyzing signals. Below we will give a classical example that shows the inefficiency of the Fourier transform in representing functions.

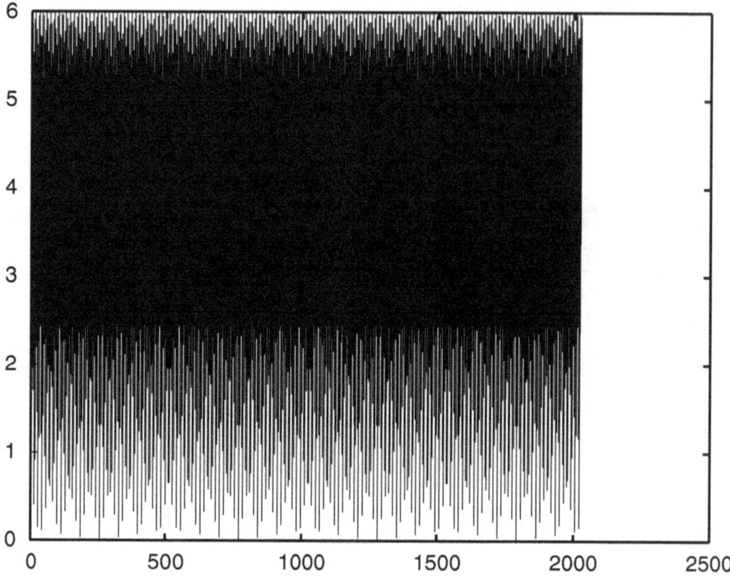

Fig. 3.3 Fourier transform of two impulses $\delta(t-0.05)$ and $\delta(t-0.42)$

Example 6 (Fourier analysis of the impulse signal). Consider the impulse "function" Dirac δ. We have

$$\int_{-\infty}^{+\infty} f(t)\delta(t)dt = f(0)\ .$$

Taking $f(t) = e^{-i2\pi st}$ in this equation we have

$$\hat{\delta}(s) = \int_{-\infty}^{+\infty} \delta(t)e^{-i2\pi st}dt = 1\ .$$

This shows that the Fourier transform of the impulse δ is a constant function of value 1. Using this result in the reconstruction Eq. (3.9), we obtain

$$\delta(t) = \int_{-\infty}^{+\infty} \hat{\delta}(f)e^{i2\pi st}ds = \int_{-\infty}^{+\infty} e^{i2\pi st}ds\ .$$

That is, exponentials of every frequency values $s \in \mathbb{R}$ must be combined in order to analyze the impulse signal.

This fact has a very simple mathematical explanation: As the Fourier transform uses modulating functions without compact support, and periodic, we must "sum" an infinite number of these functions so as to occur destructive interference. Nevertheless from the physical point of view we do not find a plausible explanation for the problem of representing such a simple signal combining an infinite number of periodic functions.

In general, if a function presents sudden changes (e.g., discontinuities), the high frequencies relative to these changes are detected by the transform, but they influence the computation of the Fourier transform along all of the domain because the modulating function does not have compact support. The Fourier analysis is therefore more efficient in the study of signals that do not suffer sudden variations along the time. These signals are called *stationary signals*.

3.4 Fourier Transform and Filtering

Even though the classical theory of filtering is developed for discrete signals, a good insight is gained by studying filtering of continuous signals. The relation of the filtering theory with the Fourier analysis is very important and will be exploited in many different parts of the book.

A filter is an operator $L: L^2(\mathbb{R}) \to L^2(\mathbb{R})$, defined on the space of signals with finite energy. An important class of filters is given by the linear, time invariant filters. A filter is linear if the operator $L: L^2(\mathbb{R}) \to L^2(\mathbb{R})$ is linear. A filter is *time invariant* if a time delay can be applied either before or after filtering, with the same result. That is,

$$L(f(t - t_0)) = (Lf)(t - t_0) .$$

Linear and space invariant filters are simple to study because they are completely determined by their values on the impulse signal $h = L\delta(t)$. Indeed, time invariance of L gives

$$L\delta(t - u) = h(t - u) .$$

Therefore,

$$L(f(t)) = L \int_{\mathbb{R}} f(u)\delta(t - u)du \tag{3.10}$$

$$= \int_{\mathbb{R}} f(u)h(t - u) = \int_{\mathbb{R}} h(u)f(t - u)du . \tag{3.11}$$

The last integral on the above equation is called the *convolution product* of h and f, and it is denoted by $h * f(t)$. That is,

$$h * f(t) = \int_{\mathbb{R}} h(u)f(t - u)du .$$

We conclude that filtering signal f with a linear, time invariant filter L is equivalent to make a convolution of f with the signal $h = L\delta(t)$. The signal h is called the *impulse response* of the filter. Sometimes h is called the *kernel* of the filter.

Applying the filter L to $e^{-i2\pi wt}$ yields

$$L(e^{-i2\pi wt}) = \int_{\mathbb{R}} h(u)e^{-i2\pi w(t-u)}du \tag{3.12}$$

$$= e^{-i2\pi wt}\int_{\mathbb{R}} h(u)e^{-i2\pi wu}du \tag{3.13}$$

$$= \hat{h}(w)e^{-i2\pi wt} . \tag{3.14}$$

This shows that each exponential $e^{-i2\pi wt}$ is an eigenvector of the filter L, and the corresponding eigenvalue is the value $\hat{h}(w)$ of Fourier transform of the impulse response function h of L.

This result is of paramount importance and it is the link between filter theory and Fourier analysis. In fact, as an immediate consequence it gives an insight in the action of the filter on the frequencies of a signal f as we will show below.

From the equation of the inverse Fourier transform we have

$$f(t) = \int_{\mathbb{R}} \hat{f}(w)e^{i2\pi wt}dw .$$

Applying L to f we get

$$L(f(t)) = \int_{\mathbb{R}} \hat{f}(w)L(e^{i2\pi wt})dw = \int_{\mathbb{R}} \hat{f}(w)\hat{h}(w)e^{i2\pi wt}dw . \tag{3.15}$$

The above equation shows that the filter L modulates the sinusoidal components $e^{i2\pi wt}$ of f, amplifying or attenuating them. Since $Lf(t) = h * f$, Eq. (3.15) can be restated as

$$F(h * f) = F(h)F(f) , \tag{3.16}$$

where F is the Fourier transform, and on the right we have a product of two functions. The Fourier transform $F(h)$ of the impulse response is called the *transfer function* of the filter. Equation (3.16) will be used to give a better insight into the filtering operation.

Linear and space invariant filters are classified according to the way they modify the frequencies of the signal f. This classification includes four basic types:

- Low-pass;
- High-pass;
- Band-pass
- Band-stop.

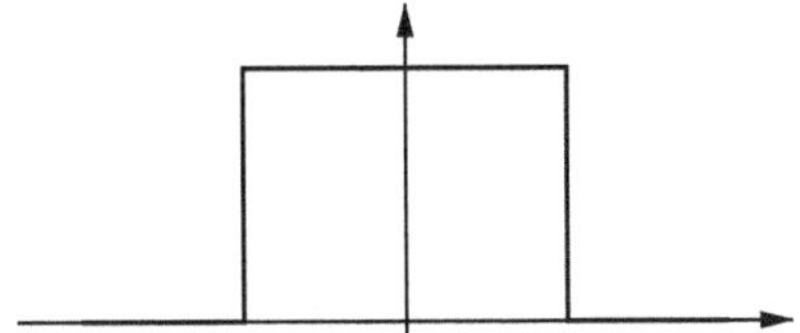

Fig. 3.4 Transfer function of an ideal low-pass filter

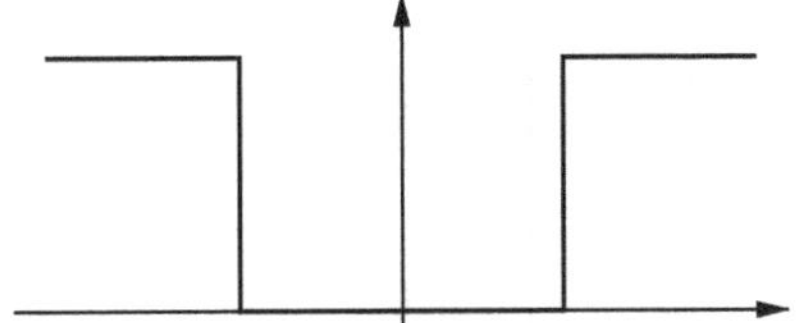

Fig. 3.5 Transfer function of an ideal high-pass filter

3.4.1 Low-pass Filters

This filter is characterized by the fact that they attenuate the high frequencies of the signal without changing the low frequencies. From Eq. (3.16) it follows that the graph of the transfer function of an ideal low-pass filter is shown in Fig. 3.4.

3.4.2 High-pass Filter

This filter has a complementary behavior to that of the low-pass filter. It attenuates the low frequencies and does not change the high frequencies. From Eq. (3.16) it follows that the graph of a transfer function of an ideal high-pass filter is shown in Fig. 3.5.

3.4.3 Band-pass Filter

This filter changes both low and high frequencies of the signal, but does not change frequencies in some interval (band) of the spectrum. The graph of the transfer function of an ideal band-pass filter is shown in Fig. 3.6. A low-pass filter is a band-pass filter for low frequencies.

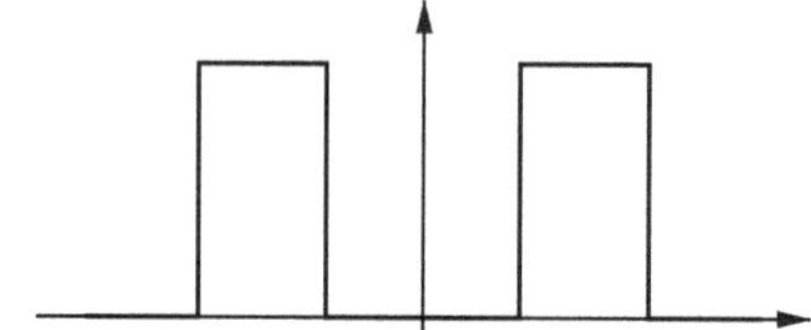

Fig. 3.6 Transfer function of an ideal band-pass filter

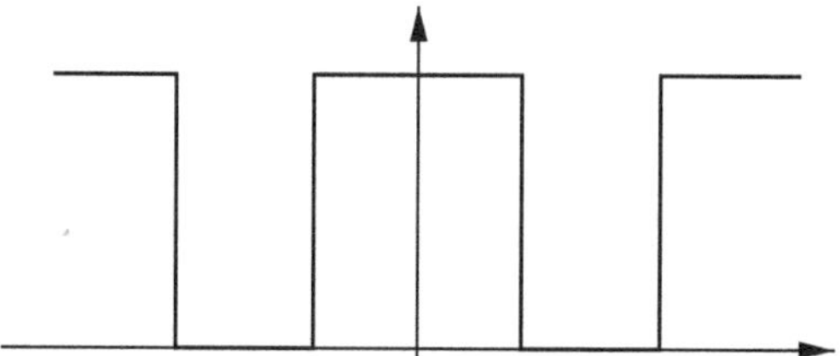

Fig. 3.7 Transfer function of an ideal band-stop filter

3.4.4 Band-stop Filter

This is the complementary of a band-pass filter. This filter affects frequencies on an interval (band) of the spectrum. Frequencies outside this frequency band are not affected. The graph of the transfer function of an ideal band-stop filter is shown in Fig. 3.7.

3.5 Fourier Transform and Function Representation

Is it possible to obtain representation and reconstruction techniques for a non-periodic function using the Fourier transform?

Equation (3.9) which defines the inverse transform gives us a clue to look for an answer to our question. In fact, this equation writes the function f as a superposition of periodical functions, "modulated" by the Fourier transform.

Certainly there exists a redundancy in this "representation" of the function f by a continuum of functions. We can eliminate this redundancy by taking only a discrete set of frequencies $s_j = \omega_0 j$, ω_0 constant, $j \in \mathbb{Z}$.

Unfortunately this discretization leads us to the Fourier series of the function f. In fact, Fourier series are particular instances of Fourier transform for discrete signals. If

$$f(t) = \sum_{n \in \mathbb{Z}} f(n)\delta(t-n) ,$$

then

$$\hat{f}(w) = \sum_{n \in \mathbb{Z}} f(n) e^{-i2\pi wn} .$$

Therefore the method that seems natural to obtain function representations from the Fourier transform is in fact inadequate. Nevertheless the Fourier theory, and its relationship with filter theory, is of great importance in the study of different representation methods. This fact will be illustrated in the next section.

3.5.1 Fourier Transform and Point Sampling

In spite of all of the weakness of the Fourier transform that we have already discussed, it is a powerful tool to analyze the problem of function representation. We will use it to understand the problem of exact reconstruction from a representation by point sampling.

An important step in this direction is to understand the spectrum of a point sampled signal. More precisely, suppose that we have a uniform sampling partition of the real numbers with interval length Δt. This length is called the *sampling period*. The discretized signal is given by

$$f_d(t) = \sum_{k \in \mathbb{Z}} f(k\Delta t)\delta(t - k\Delta t) .$$

The relation between the Fourier transform $\hat{f}$ of f and the Fourier transform $\hat{f_d}$ of the discretized signal f_d is given by

$$\hat{f_d}(s) = \frac{1}{\Delta t} \sum_{k \in \mathbb{Z}} \hat{f}\left(s - \frac{k}{\Delta t}\right) . \tag{3.17}$$

A clear interpretation of Eq. (3.17) is very important: Apart from the scaling factor $1/\Delta t$, it says that the spectrum of the sampled signal f_d is obtained from the spectrum $\hat{f}$ of the continuous signal f by translating it by multiples of $1/\Delta t$, and summing up all of the translated spectra. This is illustrated in Fig. 3.8: In (a) we show the graph of the function f; in (b) we show the graph of the Fourier transform $\hat{f}$; in (c) we show the graph of the point sampling representation (f_i) of f; in (d) we show the Fourier transform of the sampled function f_d.

It is important to remark that the translating distance varies inversely with the sampling period: The sampling period is Δt and the translating distance is $1/\Delta t$. Therefore closer samples produce more spaced translations and vice versa. Note that in particular, very high frequencies are introduced in the sampling process by the infinite translation of the spectrum.

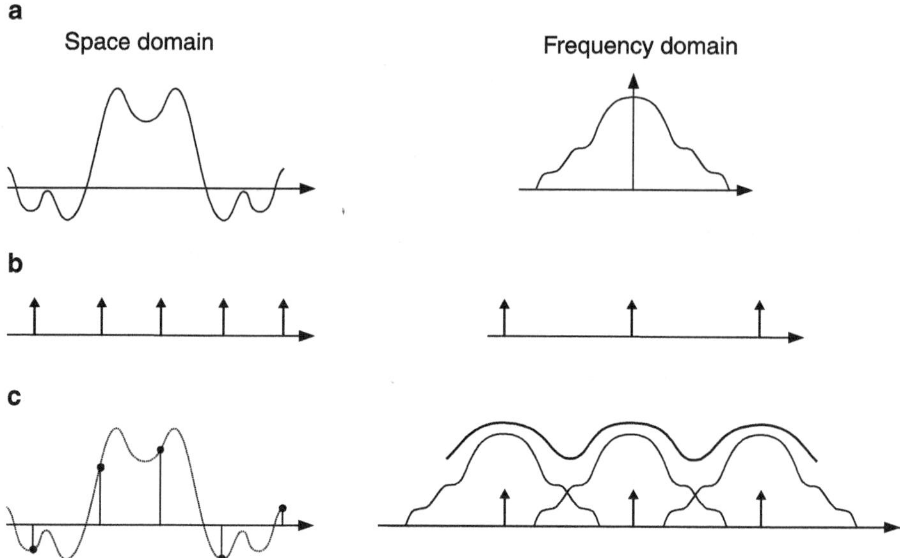

Fig. 3.8 Sampling and the Fourier spectrum

Equation (3.17) is the key point to understand the reconstruction problem when we use uniform point sampling to represent a function. This fact will be discussed in next section.

3.5.2 The Theorem of Shannon-Whittaker

In this section we will use Fourier theory to answer a question posed on Chap. 2: *Is it possible to obtain an exact representation of a function using point sampling?*

We will see that the answer is positive if we restrict the function f and at the same time impose conditions on the sampling rate. Initially we will demand the following conditions:

- The point sampling process is uniform. That is, the sampling intervals $[t_k, t_{k+1}]$ have the same length $\Delta t = t_{k+1} - t_k$.
- The Fourier transform $\hat{f}$ of the function f assumes zero values outside a bounded interval $[-\Omega, \Omega]$ of the frequency domain. We say that $\hat{f}$ has *compact support*.

Using the above restrictions, we have the classical:

Theorem 2 (Theorem of Shannon-Whittaker). *Consider a function $f: \mathbb{R} \to \mathbb{R}$ with compact support,* $\text{supp}\,(\hat{f}) \subset [-\Omega, \Omega]$, *and a uniform partition t_i, $i \in \mathbb{Z}$ of the real numbers $\mathbb{R}$ such that $2\Delta t \leq \Omega$. Then f can be reconstructed (exactly) from its point sampling representation $(f(t_i))_{i\in\mathbb{Z}}$. The reconstruction equation is given by*

$$f(t) = \sum_{k=-\infty}^{+\infty} f\left(\frac{k}{2\Omega}\right) \frac{\sin \pi(2\Omega t - k)}{\pi(2\Omega t - k)} \,. \tag{3.18}$$

The inequality $2\Delta x \leq \Omega$ is called *Nyquist limit*. It says that we must take at least 2 samples for each complete cycle of maximum frequency occurring in the function.

The result of the theorem is very intuitive. The hypothesis of the theorem says:

1. The function f should not have very high frequencies, $\text{supp}\,(\hat{f}) \subset [-\Omega, \Omega]$;
2. We take samples of f sufficiently close, such that we have at least 2 samples for a complete cycle of maximum frequency ($2\Delta t \leq \Omega$);

The conclusion of the theorem is: f can be exactly reconstructed interpolating its samples using Eq. (3.18).

A function f such that $\hat{f}$ has compact support is called a *band limited function*, because it possess frequencies within a limited interval (band) of the frequency domain. It is interesting to make a sketch of the proof of the theorem so that we can see the role of Fourier analysis and filtering.

Sketch of the Proof Consider a band limited function f. The point sampling representation transforms f into the sequence $f(t_i)$ which is null outside the points of the sampling partition, and assumes the values of f on the points of the partition.

We have seen from Eq. (3.17) that the sampling process alters the frequency spectrum of the function f, introducing very high frequencies by translating and summing up the original spectrum. In order to reconstruct f we have to recover the original frequency spectrum from the spectrum of the sampled function.

The translation distance of the spectrum of the original signal f depends on the sampling period (the length of the intervals in the sampling lattice): The smaller the period (that means more space between consecutive samples), the bigger will be the translation of the spectrum.

The Nyquist limit $2\Delta t \leq \Omega$ says that if we take the sampling period sufficiently small, the translated spectra will have disjoint domains. In this case, using an adequate low-pass filter we can recover the original spectrum of the signal f. From this original spectrum we are able to reconstruct the original signal using the inverse Fourier transform.

Details of this proof, including the computation of the reconstruction Eq. (3.18), can be found in [24].

3.6 Point Sampling and Representation by Projection

It can be shown that if we take

$$e_{\Omega}(t) = \frac{\sin \pi 2\Omega t}{\pi 2\Omega t} \,,$$

the set $\{e_\Omega(t - k\Delta t)\}$ is an orthogonal basis of the space

$$L^2_\Omega(\mathbb{R}) = \{f \in L^2(\mathbb{R}) \; ; \; \text{supp}(\hat{f}) \subset [-\Omega, \Omega]\} \; .$$

Moreover, the reconstruction Eq. (3.18) is the projection of f on this basis. In sum, the problem of point sampling and reconstruction is reduced to the problem of representation on an orthogonal basis.

If a signal f is not band-limited, that is, $f \notin L^2_\Omega(\mathbb{R})$, and it is represented using point sampling with period $\Delta t \leq \Omega/2$, and we reconstruct the sampled signal using the reconstruction Eq. (3.18) of Shannon-Whittaker, we obtain a function $\tilde{f} \in L^2(\mathbb{R})$ such that $||\tilde{f} - f||$ is minimized. In fact, $\tilde{f}$ is the orthogonal projection of f on $L^2_\Omega(\mathbb{R})$.

3.7 Point Sampling and Representation Coefficients

When we have an orthonormal basis $\{\phi_j\}$ of the space $\mathbf{L}^2(\mathbb{R})$, the representation of f in this basis is given by

$$f = \sum_n \langle f, \phi_n \rangle \phi_n \; .$$

In this case we have the representation sequence

$$f \mapsto (\langle f, \phi_n \rangle)_{n \in \mathbb{Z}} \; .$$

Note that if $\phi_n(x) = \delta(x - n)$, then

$$\langle f, \phi_n \rangle = \langle f, \delta(x - n) \rangle = f(n) \; .$$

That is, the elements of the representation sequence are samples of the function f. This fact motivates us to pose the following question:

Question 3.1. What is the relation between the elements $\langle f, \phi_n \rangle$ of the representation sequence and the samples $f(n)$ of the function f?

There is a simple answer for this question in a very particular case of great importance in the study of wavelets. We will suppose that the functions ϕ_n which constitute the orthonormal basis are obtained from a single function ϕ by translations. More precisely,

$$\phi_n(x) = \phi(x - n) \; .$$

From Parseval equation, and the identity

$$\hat{\phi}_n(s) = F(\phi(x - n)) = e^{-i2\pi ins} \hat{\phi}(s) \; ,$$

we have

$$\langle \phi_n, f\rangle = \langle \hat{\phi}_n, \hat{f}\rangle = \int_{-\infty}^{+\infty} e^{2\pi i n s}\overline{\hat{\phi}(s)}\hat{f}(s)ds = F(n) ,$$

where the function F is given by its Fourier transform

$$\hat{F}(\omega) = \overline{\hat{\phi}(\omega)}\hat{f}(\omega) . \tag{3.19}$$

Notice that

$$|\hat{F}(s)| \le ||\hat{\phi}||\ ||\hat{f}|| = ||\phi||\ ||f|| ,$$

therefore $\hat{F}(\omega)$ is integrable, and this implies that F is continuous. This shows that the values of the samples $F(n)$ are well defined.

If the function ϕ is a low-pass filter, Eq. (3.19) shows that F is obtained from the original function f by a low-pass filtering process, therefore the values of $F(n)$ are close to the values of the original function f, if it does not have great frequency variations.

For this reason, it is common to refer to the elements $\langle f, \phi_n\rangle$ from the representation sequence as *samples* of the function f, even when ϕ is not a low-pass filter. This fact is resumed in the theorem below for future references:

Theorem 3. *If $\{\phi_n = \phi(x-n)\}$ is an orthonormal basis of $\mathbf{L}^2(\mathbb{R})$, then the terms $\langle f, \phi_n\rangle$ of the representation sequence of f on the basis $\{\phi_n\}$ are obtained by filtering f, $F = f * \phi$, and sampling the resulting function F. That is,*

$$\langle f, \phi_n\rangle = \langle f, \phi(x-n)\rangle = F(n) .$$

3.8 Comments and References

There are several good books covering the theory of Fourier analysis. For a revision of the classical theory, we suggest the reader to consult [64]. This reference also covers the discrete Fourier transform. A comprehensive reference for the discrete Fourier transform, both from the conceptual and computational point of view, is found in [7].

In practice we use finite signals when implementing the operations of sampling and reconstruction on the computer. Therefore, we need to study the Fourier transform of finite signals. This Fourier transform is called *Discrete Fourier Transform* (DFT). For the reader interested in details, we recommend [24] and [7]. The reference [24] brings a chapter with a review of signal theory, adequate for those with some computer graphics background, it can be found on Chap. 1 of [24].

From the computational viewpoint an important issue related with the Fourier transform is the study of its computational complexity. That is, the study of the computational complexity of the DFT when applied to a finite signal with N samples. There are different flavors of computing with a Fast Fourier Transform, which reduces the computational complexity. For those interested in these topics, we recommend [34] and [21].

We have not stressed in this chapter a very important point when working with Fourier analysis and filtering: Since all of the theory studied here is linear, a natural representation for them is to use matrices. This approach is important specially from the computational point of view. Matrix notation is used all over in [55]. We will use some matrix notation later on this book. The reader should consult the Appendices to this book where we introduce matrix notation.

The concepts of this chapter extend naturally to real functions of several variables. Of particular importance in computer graphics is the case of functions $f: \mathbb{R}^2 \to \mathbb{R}$, which describes an image, and the case $f: \mathbb{R}^3 \to \mathbb{R}$ which is related to the study of volumetric objects. A good reference at an introductory level that covers two-dimensional signal processing is [33].

A detailed discussion of the different problems arising from incorrect reconstruction of a signal from a representation by point sampling is found in Chap. 7 of [24].

An analysis of the reconstruction problem when we use uniform point sampling representation with a sampling rate superior to the Nyquist limit is found in p. 20 of [20].

The problem of reconstructing a function from its samples called the attention of mathematicians since the beginning of the century. For details about the history of this problem, we recommend [9]. A comprehensive discussion with different versions of the problem including solutions can be found on [66].

The Theorem of Shannon-Whittaker is a very important result in the theory of function discretization. Nevertheless it presents certain deficiencies as a solution to the sampling/reconstruction problem. In particular, the hypothesis that $\hat{f}$ has compact support is too much restrictive. Several research directions are raised from the Theorem of Shannon-Whittaker. We could mention the following ones:

- Look for versions of the theorem using a weaker hypothesis than that of compact support;
- Generalize the theorem for arbitrary domains of $\mathbb{R}^n$.
- Analyze what happens with the reconstruction when the sampling is done without obeying the Nyquist limit.

Chapter 4
Windowed Fourier Transform

In this chapter we will introduce a modification in the definition of the Fourier transform in order to obtain a transform with better localization properties in the time-frequency domain. This transform will give us better results for the purposes of function representation.

4.1 A Walk in The Physical Universe

Our purpose is to obtain a transform that enables us to perform a local computation of the frequency density. The inspiration for this transform is to analyze the audio analysis performed by our auditory system. Consider for this an audio signal represented by a real function f of one variable (time).

Real time analysis The audio information we receive occurs simultaneously on time and frequency. This means that the signal f is transformed by the auditory system in a signal $\tilde{f}(t, \omega)$ that depends on the time and the frequency.

Future sounds are not analyzed This means that only values of $f(t)$ for $t \leq t_1$ can be analyzed when computing the "transform" $\tilde{f}(t, \omega)$.

The auditory system has finite memory That is, sounds that we have heard some time ago do not influence the sounds that we hear in a certain instant of time. This means that there exists a real number $t_0 > 0$ such that the computation of the "transform" $\tilde{f}(t, \omega)$ depends only on the values of t on the interval $[t - t_0, t]$.

Mathematically, the last two properties show that the modulating function used to detect frequencies in the computation of the "transform" $\tilde{f}(t, \omega)$ must have its values concentrated in a neighborhood of t. We say that it is localized in time. D. Gabor, [23] was the first to propose a transform with the above properties.

J. Gomes, L. Velho, *From Fourier Analysis to Wavelets*, IMPA Monographs 3, DOI 10.1007/978-3-319-22075-8_4

4.2 The Windowed Fourier Transform

One method to obtain a localized modulating function consists in using an auxiliary function $g(u)$ to localize the modulating function $e^{-2i\pi\omega u}$ used in the Fourier transform in a certain neighborhood of the time domain:

$$g_{\omega,t}(u) = g(u-t)e^{-2\pi i\omega u} \ . \tag{4.1}$$

This localization operator is illustrated in Fig. 4.1. We take a function $g(u)$, and for each value of $t \in \mathbb{R}$, we translate the origin to the point t and multiply the exponential by the translated function $g(u-t)$. If the function g is localized in the time domain, we obtain the desired localization of the modulating function $g(u-t)e^{-2\pi i\omega u}$.

From the above, the definition of our transform is given by

$$\begin{aligned} \tilde{f}(\omega,t) &= \int_{-\infty}^{+\infty} g(u-t)f(u)e^{-2\pi i\omega u}du \\ &= \int_{-\infty}^{+\infty} g_{t,\omega}(u)f(u)du = \langle g_{t,\omega}, f\rangle \ . \end{aligned} \tag{4.2}$$

The transform $f \mapsto \tilde{f}(t,\omega)$ is called the *windowed Fourier transform* (WFT), or the *short time Fourier transform* (STFT). Two important questions can be posed with respect to the windowed Fourier transform $\tilde{f}(t,\omega)$:

Question 4.1. Is the transform $\tilde{f}(t,\omega)$ invertible?

Question 4.2. What is the image of the transform $\tilde{f}(t,\omega)$?

The importance of these questions can be measured from the fact that the invertibility of the Fourier transform is one of its strength. The underlying idea of the importance of the inverse transform is that if we are able to obtain a good representation using the transform, its inverse will provide the reconstruction. The image of the transform is important because it measures the scope of the representable functions.

We will discuss these two questions in the following sections.

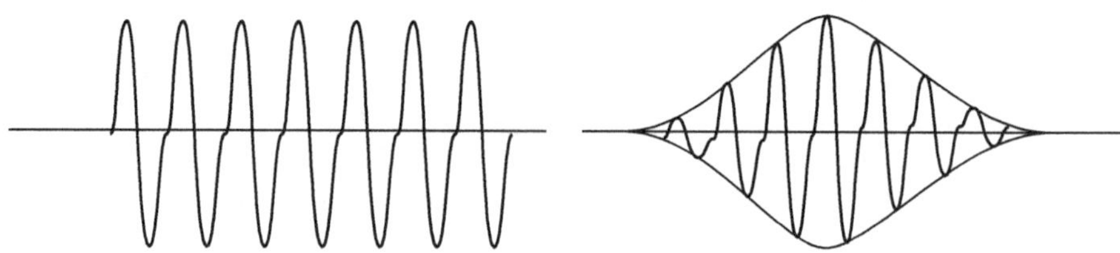

Fig. 4.1 Modulating function

4.2.1 *Invertibility of* $\tilde{f}(t, \omega)$

The problem of inverting the windowed Fourier transform consists in determining a function f from its transformed function $\tilde{f}(t, \omega)$.

From (4.2) it follows that $\tilde{f}(u, t) = \widehat{f_t}(u)$, where $f_t(u) = g(u - t)f(u)$. Applying the inverse Fourier transform, we have

$$g(u - t)f(u) = f_t(u) = \int_{-\infty}^{+\infty} \tilde{f}(\omega, t)e^{2\pi i\omega u}d\omega \ .$$

We cannot divide by $g(u-t)$ to get $f(u)$, because the function g might as well assume zero values. Multiplying both members of the equation $g(u - t)$, and integrating in t we obtain

$$\int_{-\infty}^{+\infty} |g(u - t)|^2 f(u)dt = \int_{-\infty}^{+\infty} \int_{-\infty}^{+\infty} e^{-2\pi i\omega u}g(u - t)\tilde{f}(\omega, t)d\omega dt \ .$$

That is,

$$f(u) = \frac{1}{||g||^2} \iint_{\omega,t} g(u - t)e^{2\pi i\omega u}\tilde{f}(\omega, t)d\omega \ . \tag{4.3}$$

As we did for the Fourier transform, Eq. (4.3) can be interpreted into two different ways:

1. It is an equation to compute f from its windowed Fourier transform $\tilde{f}(\omega, t)$;
2. It decomposes the function f as an infinite superposition of localized waveforms

$$g_{\omega,t}(u) = g(u - t)e^{2\pi i\omega u}.$$

These waveforms are called *time-frequency atoms.*

4.2.2 *Image of the Windowed Fourier Transform*

In this section we will discuss the second question posed before: *What is the image space of the windowed Fourier transform?* We will only give some comments concerning the answer.

Given a function f its windowed Fourier transform $\tilde{f}(\omega, t)$ is a function of two variables. It is possible to prove that if $f \in \mathbf{L}^2(\mathbb{R})$, then $\tilde{f}(\omega, t)\mathbf{L}^2(\mathbb{R}^2)$. Also, it is possible to show that the image of the transform $\tilde{f}$ does not cover the whole space $\mathbf{L}^2(\mathbb{R}^2)$. Therefore, the posed question consists in characterizing the image set of the windowed Fourier transform. We will not solve this problem here. A solution to it can be found in [31], p. 56, or Theorem 4.1 of [37].

It is important to compare the result here with the analogous result for the Fourier transform: The Fourier transform is an isometry of the space $L^2(\mathbb{R})$. In particular its image is the whole space.

4.2.3 *WFT and Function Representation*

Since our main focus is the problem of function representation, a natural question to be posed now would be:

Question 4.3. Is it possible to obtain a function representation using the windowed Fourier transform?

In fact since the windowed Fourier transform has good localization properties in the time-frequency domain, we should expect that the discretization of Eq. (4.3) would give good discrete time-frequency atomic representations.

4.3 Time-frequency Domain

We have noticed that if $f \in \mathbf{L}^2(\mathbb{R})$, then $\tilde{f}(\omega, t) \in \mathbf{L}^2(\mathbb{R}^2)$. Therefore the windowed Fourier transform of a function is defined on the domain (ω, t), called the *time-frequency* domain.

From the definition of the windowed Fourier transform (4.2) we know that if g is well localized in time (i.e., g is small outside of a small time interval), then $\tilde{f}$ is also well localized in time. How about the frequency localization of $\tilde{f}$?

From Eq. (4.2) which defines the windowed Fourier transform, we have

$$\tilde{f}(\omega, t) = \langle g_{\omega,t}, t\rangle = \langle \hat{g}_{\omega,t}, \hat{f}\rangle ,$$

where the second equation follows from Parseval's identity. We conclude that if g has good localization properties in the frequency domain (i.e., $\hat{g}$ is small outside an interval of frequency ω), then the transform $\tilde{f}$ is also localized in frequency.

Therefore the windowed Fourier transform enables us to analyze the function f in the time-frequency domain, in the sense that we have localized information both in time and frequency domain. This result is completely in accordance with the problem we have discussed before: Detect frequencies of the function f, and localize them on the time domain.

How precisely can we localize the information about f in the time-frequency domain? An answer to this question is given below.

4.3.1 The Uncertainty Principle

From the previous section we could conclude that a finer analysis of a function $f \in L^2(\mathbb{R})$ could be obtained by using window functions g with very good localization properties on the time-frequency domain (ω, t).

Unfortunately there is a limit to the localization precision in the time-frequency domain. This limitation comes from a general principle that governs the time-frequency transforms. This is the *uncertainty principle* which will be discussed now. In simple terms the statement of this principle is: *We can not obtain precise localization simultaneously in the time and frequency domains*. The intuition behind this principle is simple: To measure frequencies we must observe the signal for some period of time. The more precision we need in the frequency measurements the larger the time interval we have to observe.

In order to give a more quantitative statement of the uncertainty principle, we have to define precisely the notion of "information localization" of a function. For this, we will suppose that the norm $\mathbf{L}^2(\mathbb{R})$ of the window function g is 1, that is, $||g||^2 = 1$. It follows from the equation of Plancherel that $||\hat{g}||^2 = 1$. We may consider g and $\hat{g}$ as probability distributions, then the averages of g and $\hat{g}$ are computed by

$$t_0 = \int_{-\infty}^{+\infty} t|g(t)|dt, \quad \text{and} \quad \omega_0 = \int_{-\infty}^{+\infty} \omega|\hat{g}(\omega)|d\omega ,$$

respectively. The size of the localization interval of g and $\hat{g}$ is given the standard deviation

$$T^2 = \int_{-\infty}^{+\infty} (t - t_0)^2 |g(t)|^2 dt ,$$

and

$$\Omega^2 = \int_{-\infty}^{+\infty} (\omega - \omega_0)^2 |\hat{g}(\omega)|^2 d\omega .$$

With the above definitions, the uncertainty principle states that

$$4\pi \Omega T \geq 1 .$$

Note that if g is well localized in frequency (Ω small) then $T \geq 1/4\pi\Omega$ cannot be small, therefore g does not have good localization in time. The same reasoning applies for frequency localization.

The localization of the signal in the time-frequency domain (ω, t) is represented geometrically by the rectangle of dimensions $T \times \Omega$. This rectangle is called *uncertainty window* or *information cell* of the transform. From the uncertainty principle, the area of this rectangle is $\geq 1/4\pi$ (see Fig. 4.2).

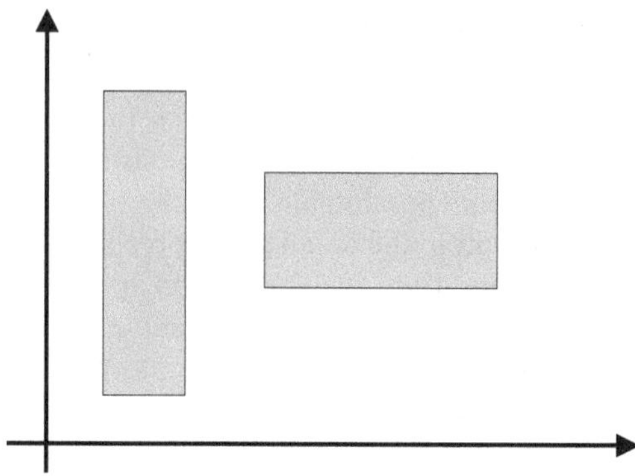

Fig. 4.2 Information cells

The uncertainty principle shows that do not exist point coordinates in the time-frequency domain. Or, to state this more precisely, the coordinates in the time-frequency domain are the centroids of density distributions, and don't have great importance to the analysis of functions, unless for the fact that they are used to measure the dimensions of the uncertainty windows.

4.4 Atomic Decomposition

Our main goal is to provide better representations of a signal. The purpose of introducing the windowed Fourier transform is to obtain representations in the time-frequency domain. More precisely, given a function f we must obtain a decomposition

$$f = \sum_{\lambda \in \Omega} a_\lambda g_\lambda \,, \tag{4.4}$$

where Ω is discrete, and the functions g_λ have good localization properties in the time-frequency domain. The functions g_λ are called *time-frequency atoms*, and the reconstruction Eq. (4.4) is called *atomic decomposition*.

The atomic decomposition defines a representation operator

$$R(f) = (a_\lambda)_{\lambda \in \Omega} \in \ell^2 \,.$$

Each atom in this representation constitutes a basic element used to measure the frequency density of the function in a small period of time. Each of these atoms is represented by a rectangle whose sides indicate the measure of localization according to the uncertainty principle. The degree of uncertainty in time and in

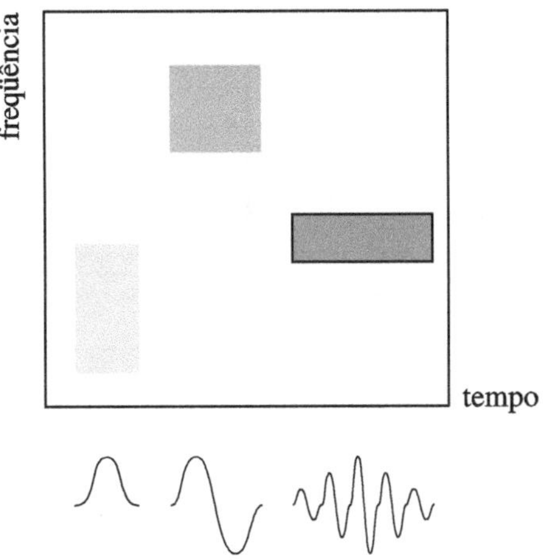

Fig. 4.3 Atoms in the time-frequency domain

frequency is indicated by the width and height of the rectangle. The localization of the atom in the time-frequency domain is given by the coordinates of the center of the rectangle, or by the coordinates of some of its vertices. From the uncertainty principle the area of each rectangle is $\geq 1/4\pi$.

We associate a gray color with each atom to indicate its energy in the decomposition. The energy is directly related with the value of the coefficients in the reconstruction equation. In Fig. 4.3 we depict some atoms. The associated signal to the atom on the left presents small localization in frequencies and has small energy; the central atom has better localization of frequencies (complete cycle) and therefore has more energy; the atom to the right has a good frequency localization (several cycles are encompassed) and a high energy.

Given a signal f represented by some finite atomic decomposition

$$f_N(t) = \sum_{k=0}^{N-1} a_N(k)\phi_{N,k}(t) ,$$

this representation depicted by the corresponding rectangles of each atom $\phi_{N,k}$, and the corresponding energy component $a_N(k)\phi_{N,k}$.

In the case of representation by uniform sampling, we have an exact localization in time and no localization in the frequency domain. The representation in the time-frequency domain is illustrated in Fig. 4.4(a), where the distribution between the vertical segments is given by the sampling interval.

The atoms of the discrete Fourier transform

$$1, e^{\frac{2\pi i}{N}}, \dots, e^{\frac{2\pi i(N-1)s}{N}}$$

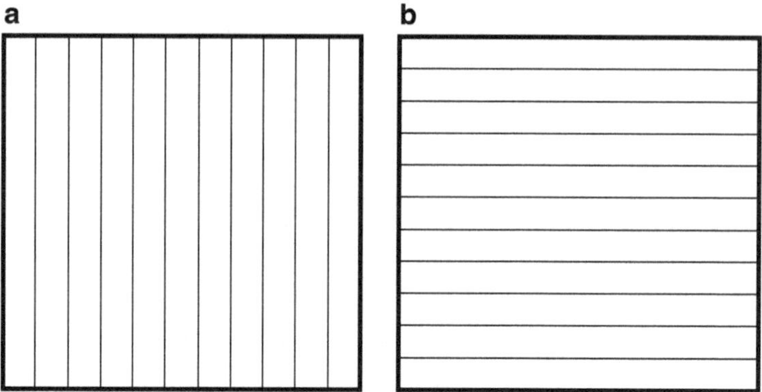

Fig. 4.4 **a** Point sampling. **b** Fourier sampling

constitute an orthonormal basis. We have N discrete values of frequency $0, s, 2s, \ldots, (N-1)s$. These atoms have no time localization, therefore the atomic representation of a function using this basis is localized in the frequency domain and has no time localization. This fact is depicted in Fig. 4.4(b).

4.5 WFT and Atomic Decomposition

The natural way to obtain atomic decompositions of the function f using the windowed Fourier transform would be using the inverse transform (4.3) which writes f as a superposition of indexed functions by the time frequency parameters. This equation gives us an indication that by discretizing the time and frequency parameters we obtain representation/reconstruction methods for functions $f \in \mathbf{L}^2(\mathbb{R})$.

In order to achieve this we should look for discrete versions of the windowed Fourier transform. We fix t_0 and ω_0 and we take discrete values of time $t = nt_0$, and discrete values of frequency $\omega = m\omega_0$, $n, m \in \mathbb{Z}$. In the time-frequency domain we have the uniform lattice

$$\Delta_{t_0,\omega_0} = \{(mt_0, n\omega_0)\ ;\ m, n \in \mathbb{Z}\}\ ,$$

depicted in Fig. 4.5.

In this case we write the transform $\tilde{f}$ in the form

$$\tilde{f}_{m,n} = \int_{-\infty}^{+\infty} f(u)g(u - nt_0)e^{-2\pi i m\omega_0 u}du\ , \tag{4.5}$$

which is called *discrete windowed Fourier transform* (DWFT).

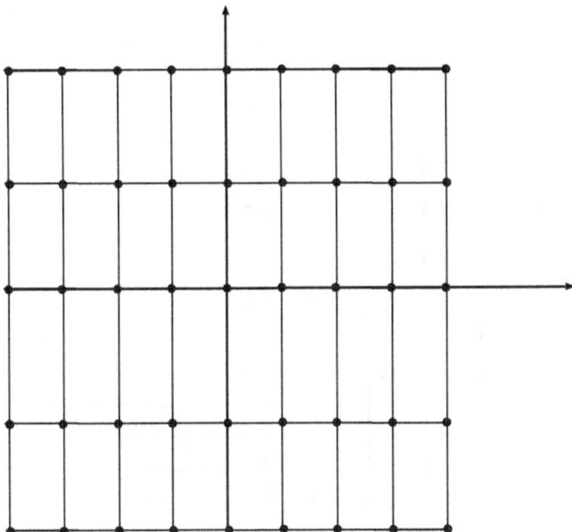

Fig. 4.5 Uniform lattice in time-frequency domain

The discrete windowed Fourier transform uses a countable family of time frequency atoms

$$g_{m,n}(u) = e^{i2\pi m\omega_0 u} g(u - nt_0) \ .$$

If this family constitute a frame, we can obtain a representation

$$f \mapsto (\langle f, g_{m,n} \rangle)_{m,n \in \mathbb{Z}} \ ,$$

of the function f. Moreover, f can be reconstructed from this representation using the reciprocal frame $\widetilde{g_{m,n}}$, as described in Chap. 2:

$$f = \sum_{m,n} \langle f, g_{m,n} \rangle \widetilde{g_{m,n}} \ .$$

Geometrically, this result means that when we position the information cells of the time-frequency atoms $g_{m,n}$ on the vertices of the lattice, we cover the whole time-frequency plane. This fact is illustrated in Fig. 4.6 for two distinct discretizations of the domain.

Now the crucial question to complement the above results is:

Question 4.4. Does there exist frames $g_{m,n}$ with good localization properties in the time-frequency plane?

A detailed analysis of this question, with several related references, can be found in Chap. 4 of [20]. We will give a briefing of the results here:

1. If $t_0\omega_0 > 2\pi$, then frames $g_{m,n}$ do not exist.

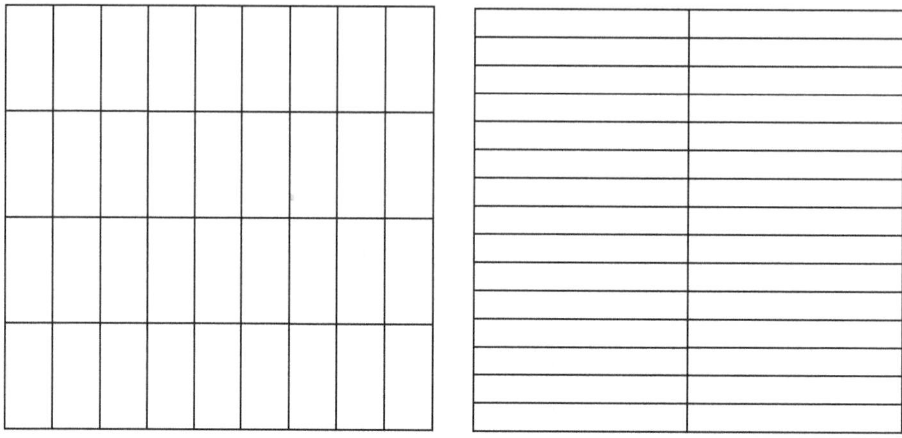

Fig. 4.6 Decompositions of the time-frequency domain using the DWFT

2. If $t_0\omega_0 = 2\pi$, frames do exist but they do not have good localization properties in the time-frequency domain. In fact, if we take

$$g(x) = \begin{cases} 1 & \text{if } x \in [0,1] \\ 0 & \text{if } x < 0 \text{ or } x > 1 \end{cases},$$

then $g_{m,n}(x)$ is a basis. The same happens if we take $g(x) = \sin \pi x/\pi x$.
3. If $t_0\omega_0 < 2\pi$, then there exists tight frames with good time-frequency localization properties. A construction of such a tight frame is found in [19].

From the point of view of time-frequency, the atomic decomposition of a function using the discrete windowed Fourier transform gives a uniform decomposition in rectangles, according to the illustration in Fig. 4.6. We will give some examples.

Example 7 (Sines with impulses). Consider the signal defined by the function

$$f(t) = \sin(2\pi 516.12t) + \sin(2\pi 2967.74t) + \delta(t - 0.05) + \delta(t - 0.42) .$$

This signal consists of a sum of two senoids with frequencies 516.12 Hz and 2967.74 Hz, with two impulses of order 3 for time values of $t_0 = 0.05$ s and $t_1 = 0.42$ s. The graph of the signal is shown in the image on the left of Fig. 4.7. The graph of its Fourier transform is depicted in the image on the right.

The analysis of this signal using the Fourier transform was done in Chap. 2. Our goal here is to revisit this analysis using the windowed Fourier transform. For this, we use a Hamming window with different sizes. A Hamming window is defined by a cosine curve $a + b\cos t$ for convenient values of a and b. Its graph is shown in Fig. 4.8.

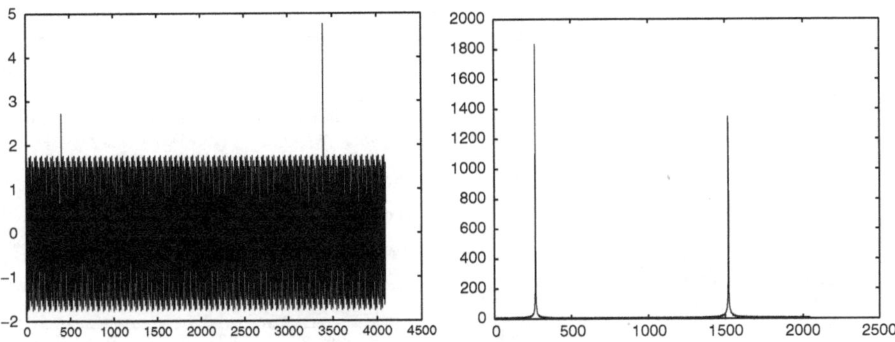

Fig. 4.7 Signal and its Fourier transform

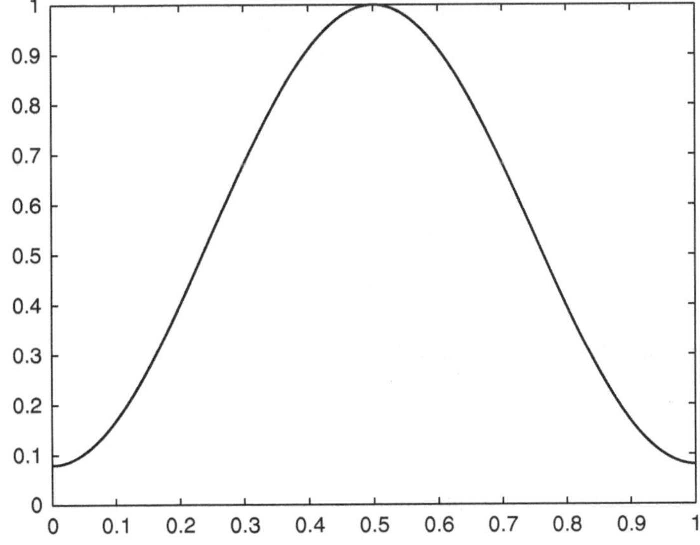

Fig. 4.8 Hamming window

The images (a), (b), and (c) of Fig. 4.9 show the information cells of the atomic decomposition of the signal f using the discrete windowed Fourier transform. In the decomposition shown in (a) we have used a Hamming window of width 32 (i.e., 32 samples) in the decomposition shown in (b) we used a Hamming window of width 64 (64 samples), and in (c) we have used a Hamming window of width 256.

An analysis of the decompositions in the figure shows us the following:

- In (a) we see that two impulses were detected with a good localization in time. The two sine waves also have been detected but the localization of their frequencies is not good. Moreover several information cells show up which are not directly related with the dominating frequencies of the signal.

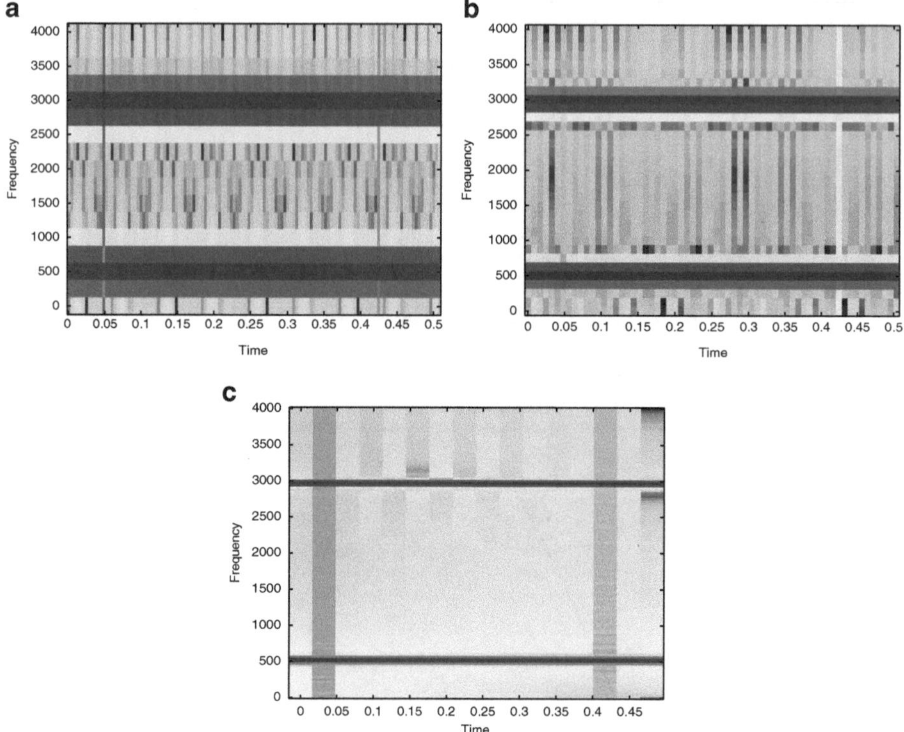

Fig. 4.9 Atomic decomposition using windowed Fourier transform

- In (b) we have a better localization of the sine frequencies, but the information cells are not able to distinguish the two impulses.
- In (c) we have a good localization of the two impulses in time and good localization of the sinusoidal frequencies. Also it should be noticed that most of the information cells in the figure are related with the relevant frequencies of the signal. Nevertheless, it should be remarked that if the two pulses were close together, we would not be able to distinguish them.

The above example shows that making an adequate adjustment of the size of the window, we can obtain satisfactory results in the analysis of the signal using the windowed Fourier transform. The results are much better than those of the analysis using the Fourier transform in the previous chapter.

Nevertheless we must take into consideration that the signal that we have used is very simple: it contains only two distinguished frequencies of the sine waves, and the two impulses.

Now we will give an example of a signal which is very difficult to be analyzed using the windowed Fourier transform. The idea of constructing such an example is very simple: The information cells of the windowed Fourier transform have constant

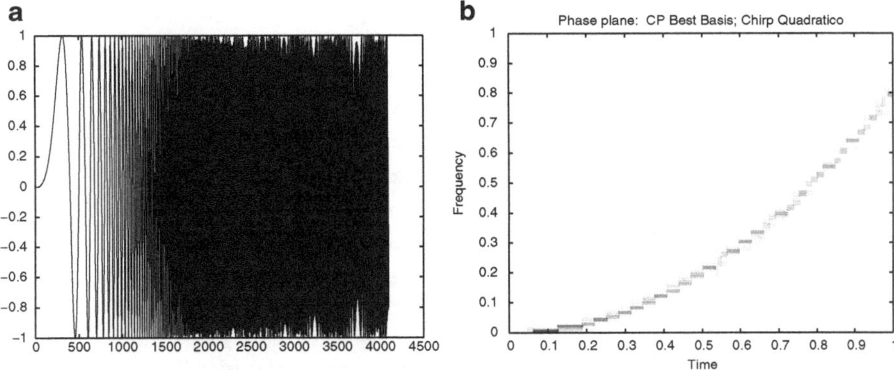

Fig. 4.10 **a** Quadratic chirp signal. **b** Ideal atomic decomposition of the signal

width, thus if a signal has frequencies which varies between different orders of magnitude, it is very difficult to obtain an adequate width of a window that is able to detect all of the frequency values. We will give an example illustrating this fact.

Example 8 (Quadratic chirp signal). Consider the signal defined by the function

$$f(t) = \sin^2(t^2).$$

The frequencies of this signal have a quadratic growth along the time. The graph of the function f is shown in Fig. 4.10(a). Since the signal has low frequencies close to the origin $t = 0$ and they increase arbitrarily as the time increases, we conclude that a good atomic decomposition of this signal should have time-frequency atoms as illustrated in Fig. 4.10(b).

Figure 4.11 shows two atomic decompositions of the signal f using the windowed Fourier transform. In both we have used Hamming windows of different sizes. It should be noticed that the relevant frequencies are detected and are correlated along a parabola, as we predicted. Nevertheless several other information cells appear that are not related with relevant frequencies of the signal. Moreover the information cells corresponding to the frequencies of the signal do not possess good localization properties.

The above example makes explicit the limitation of the windowed Fourier transform: It uses a fixed window size. In the next chapter we will introduce the wavelet transform in an attempt to fix this problem. The problem of representation and reconstruction of signals will be revisited with this transform. As we will see, better atomic decompositions will be obtained.

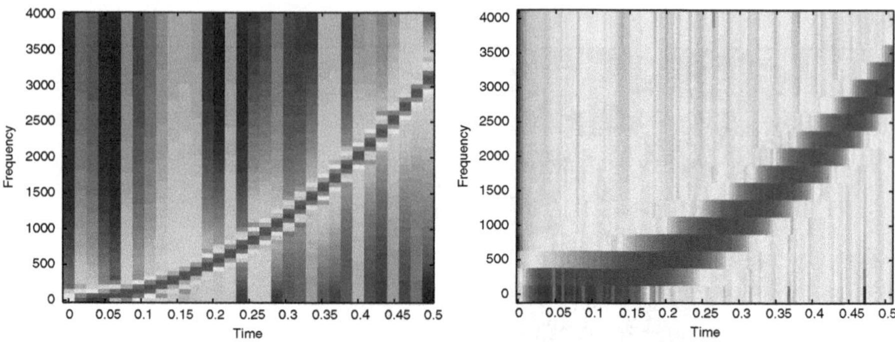

Fig. 4.11 Time-frequency decomposition of chirp signal using DWFT

4.6 Comments and References

The windowed Fourier transform was introduced in the literature by D. Gabor [23]. D. Gabor used Gaussian windows, for this reason the windowed Fourier transform with gaussian windows is also called *Gabor transforms* in the literature.

The reference [31] contains useful information with detailed proofs, the only drawback is that it requires more familiarity of the reader with techniques from functional analysis and measure theory. In particular it generalizes the concept of frames for non-discrete frames, and along with the equations of the resolution of the identity provides elegant proofs of several results.

The study of the windowed Fourier transform in [20] is quite complete. Nevertheless frames for the windowed Fourier transform are discussed along with the problem of wavelet frames. This could bring some difficulties for those with less experience.

Chapter 5
The Wavelet Transform

In this chapter we will introduce the wavelet transform with the purpose of obtaining better representation of functions using atomic decompositions in the time-frequency domain.

5.1 The Wavelet Transform

The windowed Fourier transform introduces a scale (the width of the window), and analyzes the signal from the point of view of that scale. If the signal has important frequency details outside of the scale, we will have problems in the signal analysis:

- If the signal details are much smaller than the width of the window, we will have a problem similar to the one we faced with the Fourier transform: The details will be detected but the transform will not localize them.
- If the signal details are larger than the width of the window, they will not be detected properly.

To solve this problem when we analyze a signal using the windowed Fourier transform, we must define a transform which is independent of scale. This transform should not use a fixed scale, but a variable one.

The scale is defined by the width of the modulation function. Therefore we must use a modulation function which does not have a fixed width. Moreover the function must have good time localization. To achieve this we start from a function $\psi(t)$ as a candidate of a modulation function, and we obtain a family of functions from ψ by varying the scale: We fix $p \geq 0$ and for all $s \in \mathbb{R}$, $s \neq 0$, we define

$$\psi_s(u) = |s|^{-p}\psi\left(\frac{u}{s}\right) = \frac{1}{|s|^p}\psi\left(\frac{u}{s}\right) . \tag{5.1}$$

J. Gomes, L. Velho, *From Fourier Analysis to Wavelets*,
IMPA Monographs 3, DOI 10.1007/978-3-319-22075-8_5

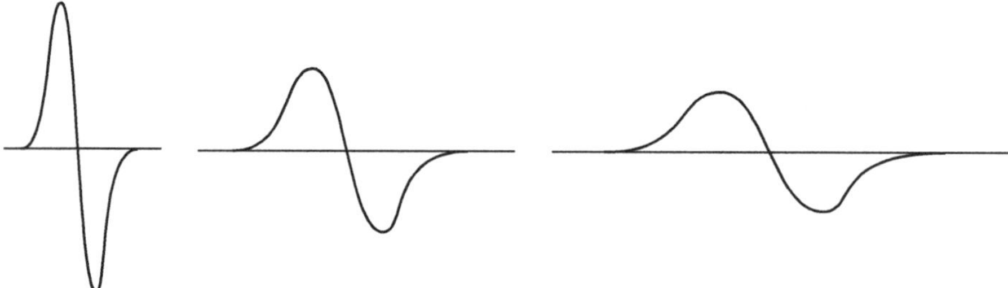

Fig. 5.1 Scales of a function: **a** $s < 1$; **b** $s = 1$; **c** $s > 1$

If ψ has width T (given as the standard deviation as explained in Chap. 3), then the width of ψ_s is sT. The modulation of the function ψ by the factor $1/|s|^2$ increases its amplitude when the scale s decreases and vice versa. In terms of frequencies, we can state: For small scales s, ψ_s has high frequencies, and as s increases the frequency of ψ_s decreases. This fact is illustrated in Fig. 5.1.

Analogous to what we did with the windowed Fourier transform of a function, we need to localize each function ψ_s in time. For this we define for each $t \in \mathbb{R}$ the function

$$\psi_{s,t}(u) = \psi_s(u-t) = |s|^{-p}\psi\left(\frac{u-t}{s}\right) = \frac{1}{|s|^p}\psi\left(\frac{u-t}{s}\right) . \tag{5.2}$$

Note that if $\psi \in \mathbf{L}^2(\mathbb{R})$, then $\psi_{s,t} \in \mathbf{L}^2(\mathbb{R})$, and

$$||\psi_{s,t}||^2 = |s|^{1-2p}||\psi||^2 .$$

By taking $p = 1/2$, we have $||\psi_{s,t}|| = ||\psi||$.

Now we can define a transform on $\mathbf{L}^2(\mathbb{R})$ in a similar way that we defined the windowed Fourier transform, using functions from the family $\psi_{s,t}$ as modulating functions. More precisely, we have

$$\tilde{f}(s,t) = \int_{-\infty}^{+\infty} f(u)\psi_{s,t}(u)du = \langle \psi_{s,t}, f \rangle . \tag{5.3}$$

This transform is known by the name of the *wavelet transform*.

As we did for the windowed Fourier transform, we can pose the following questions concerning the wavelet transform:

Question 5.1. Is the wavelet transform $\tilde{f}(s,t)$ invertible?

Question 5.2. What is the image of the wavelet transform $\tilde{f}(s,t)$?

In the previous chapter we have explained the importance of these two questions for function representation using time-frequency atoms.

5.1.1 *Inverse of the Wavelet Transform*

By definition we have

$$\tilde{f}(s,t) = \langle \psi_{s,t}, f \rangle = \langle \hat{\psi}_{s,t}, \hat{f} \rangle.$$

Moreover,

$$\hat{\psi}_{s,t}(\omega) = |s|^{1-p} e^{-2\pi i \omega t} \hat{\psi}(s\omega). \tag{5.4}$$

From this it follows that

$$\tilde{f}(s,t) = |s|^{1-p} \int_{-\infty}^{+\infty} e^{2\pi i \omega t} \hat{\psi}(s,\omega) \hat{f}(\omega) d\omega \tag{5.5}$$

$$= |s|^{1-p} F^{-1} \left(\hat{\psi}(s\omega) \hat{f}(\omega) \right), \tag{5.6}$$

where F indicated the Fourier transform.

Applying the Fourier transform to both sides of the equation we obtain

$$\int_{-\infty}^{+\infty} e^{-2\pi i \omega t} \tilde{f}(s,t) dt = |s|^{1-p} \hat{\psi}(s\omega) \hat{f}(\omega). \tag{5.7}$$

From the knowledge of $\hat{f}$ we can obtain f using the inverse transform. But we cannot simply divide the above equation by $\hat{\psi}$, because it might have zero values. Multiplying both sides of (5.7) by $\hat{\psi}(s\omega)$, and making some computations we obtain the result below:

Theorem 4. *If ψ satisfies the condition*

$$C = \int_{-\infty}^{+\infty} \frac{|\hat{\psi}(u)|^2}{|u|} < \infty, \tag{5.8}$$

then

$$f(u) = \frac{1}{C} \iint_{\mathbb{R}^2} |s|^{2p-3} \tilde{f}_{s,t}(u) \psi_{s,t}(u) ds dt. \tag{5.9}$$

This theorem answers the first question posed at the end of the previous section: The wavelet transform is invertible and Eq. (5.9) reconstructs f from its wavelet transform.

As we did with the windowed Fourier transform, we can read Eq. (5.9) of the inverse wavelet transform in two distinct ways:

1. The function f can be recovered from its wavelet transform;
2. The function f can be decomposed as a superposition of the time-frequency atoms $\psi_{s,t}(u)$.

We have seen that the second interpretation is of great importance because, as in the case of the windowed Fourier transform, it will lead us to obtain good representations by atomic decompositions of the function f.

5.1.2 *Image of the Wavelet Transform*

In this section we will discuss the second question we asked before about the image of the wavelet transform.

The wavelet transform, similarly with the windowed Fourier transform, takes a function $f \in \mathbf{L}^2(\mathbb{R})$ into a function $\tilde{f}(s,t)$ of two variables. A natural question consists in computing the image of the transform.

The interested reader should consult [31], p. 69. Besides characterizing the image space, this reference brings a proof that the wavelet transform defines an isometry over its image. We will not go into details of the computation here.

5.2 Filtering and the Wavelet Transform

Equation (5.3) that defines the wavelet transform can be written as a convolution product

$$\tilde{f}(s,t) = f * \psi_s(u) \ ,$$

where $\psi_s(u)$ is defined in (5.1). Thus the wavelet transform is a linear space-invariant filter. In this section we will discuss some properties of the wavelet filter.

The condition (5.8) that appears in the hypothesis of the Theorem 4 is called *admissibility condition*. A function ψ that satisfies this condition is called a *wavelet*.

From the admissibility condition it follows that

$$\lim_{u \to 0} \hat{\psi}(u) = 0 \ .$$

If $\hat{\psi}(u)$ is continuous, then $\hat{\psi}(0) = 0$, that is,

$$\int_{-\infty}^{+\infty} \psi(u) du = 0 \ .$$

Geometrically, this condition states that the graph of the function ψ must oscillate so as to cancel positive and negative areas in order to have integral zero. Therefore the graph of ψ has the form of a wave. In fact since ψ should have good time localization properties it has a form of a "small wave" (see Fig. 5.2). That is why ψ is named by *wavelet*.

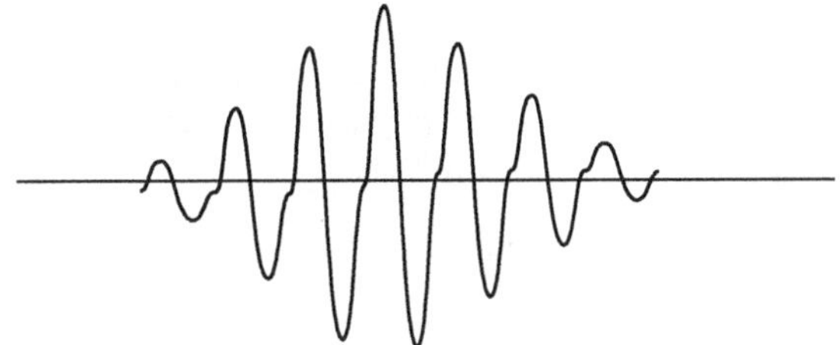

Fig. 5.2 Graph of a wavelet

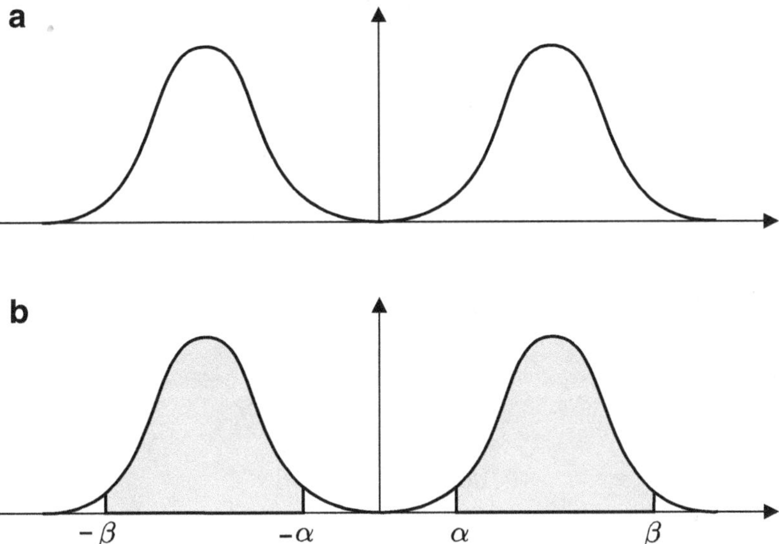

Fig. 5.3 Fourier transform of a wavelet

Another important conclusion can be drawn from the above computations. Since $\hat{\psi}(u) \in \mathbf{L}^2(\mathbb{R})$, then

$$\lim_{u \to 0} \hat{\psi}(u) = 0 .$$

Along with the fact that $\hat{\psi}(0) = 0$, we conclude that the graph of the Fourier transform $\hat{\psi}$ is as depicted in Fig. 5.3(a).

If $\hat{\psi}$ has a fast decay when $u \to 0$ and $u \to \infty$, then $\hat{\psi}(u)$ is small outside of a small frequency band $\alpha \leq |u| \leq \beta$ (see Fig. 5.3(b)). It follows from Eq. (5.4) that $\hat{\psi}_{s,t} \approx 0$ outside of the frequency band

$$\frac{\alpha}{|s|} \leq |u| \leq \frac{\beta}{|s|}.$$

Moreover, from Eq. (5.6) the wavelet transform $\tilde{f}$ does not contain information about f outside of this spectrum interval. In sum, the computations above show that "*the wavelet transform is a linear, time invariant band-pass filter.*"

The next two examples are taken from [31].

Example 9 (Blur Derivative). Consider a function ϕ of class C^∞, satisfying the conditions

$$\phi \geq 0;$$

$$\int_{\mathbb{R}} \phi(u)du = 1;$$

$$\int_{\mathbb{R}} u\phi(u)du = 0;$$

$$\int_{\mathbb{R}} u^2\phi(u)du = 1.$$

That is, ϕ is a probability distribution with average 0 and variance (width) 1. Suppose that

$$\lim_{u\to+\infty} \frac{\partial^{n-1}\phi}{\partial u^{n-1}}(u) = 0 \ .$$

Defining

$$\psi^n(u) = (-1)^n \frac{\partial^n \phi}{\partial u^n}(u) \ ,$$

we have

$$\int_{\mathbb{R}} \psi^n(u)du = 0 \ .$$

That is, ψ^n satisfies the admissibility condition (5.8). Therefore we can define a wavelet transform

$$\tilde{f}(s,t) = \int_{\mathbb{R}} \psi^n_{s,t}(u) f(u)du \ , \tag{5.10}$$

where

$$\psi^n_{s,t}(u) = \frac{1}{s}\psi^n\left(\frac{u-t}{s}\right) \ .$$

(We are taking $p = 1$ in Eq. (5.2) that defines $\psi_{s,t}(u)$). In an analogous way we define

$$\phi_{s,t}(u) = \frac{1}{s}\phi\left(\frac{u-t}{s}\right) \ .$$

From the definition of ψ^n we have that

$$\psi_{s,t}^{-n}(u) = (-1)^n s^{-n} \frac{\partial^n \phi_{s,t}}{\partial u^n}(u) = s^{-n} \frac{\partial^n \phi_{s,t}}{\partial t^n}(u) . \tag{5.11}$$

From Eqs. (5.10) and (5.11) it follows that

$$\tilde{f}(s,t) = s^{-n} \frac{\partial^n}{\partial t^n} \int_{\mathbb{R}} \phi_{s,t}(u) f(u) du. \tag{5.12}$$

The above integral is a convolution product of the function f with the function $\phi_{s,t}$, therefore it represents a low-pass filtering linear time-invariant filtering operation of the function f, which is dependent on the scale s. We will denote this integral by $\overline{f}(s,t)$. Therefore we have

$$\tilde{f}(s,t) = s^{-n} \frac{\partial^n \overline{f}(s,t)}{\partial t^n}, \tag{5.13}$$

that is, the wavelet transform of f is the n-th time derivative of the average of the function f on scale s. This derivative is known in the literature by the name of *blur derivative*.

We know that the n-th derivative of f measures the details of f in the scale of its definition. Therefore, Eq. (5.13) shows that the wavelet transform $\tilde{f}(s,t)$ gives the detail of order n of the function f, in the scale s. Keeping this wavelet interpretation in mind is useful, even when the wavelet does not come from a probability distribution.

Example 10 (The Sombrero Wavelet). We will use a particular case of the previous example to define a wavelet transform. Consider the Gaussian distribution

$$\phi(u) = \frac{1}{\sqrt{2\pi}} e^{-u^2/2} ,$$

with average 0 and variance 1. The graph of this function is depicted in the image on the left of Fig. 5.4. Using the notation of the previous example, we have

$$\psi^1(u) = -\phi'(u) = \frac{1}{\sqrt{2\pi}} u e^{-u^2/2} ,$$

and

$$\psi^2(u) = \phi''(u) = \frac{1}{\sqrt{2\pi}} (u^2 - 1) e^{-u^2/2} .$$

The function $-\psi^2$ is known as the "sombrero" function, because of the shape of its graph, shown in the right of Fig. 5.4.

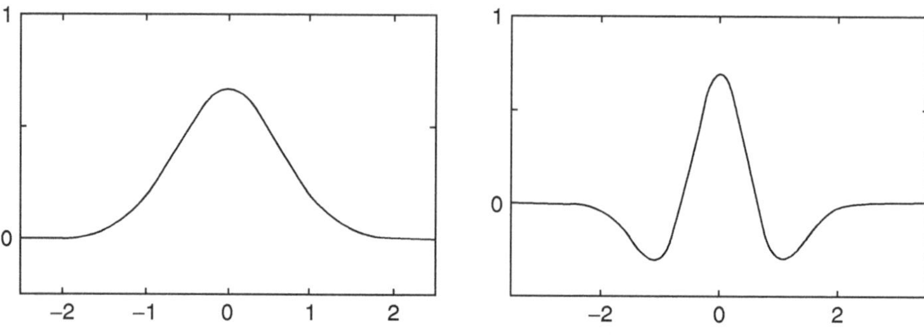

Fig. 5.4 Graph of the sombrero wavelet

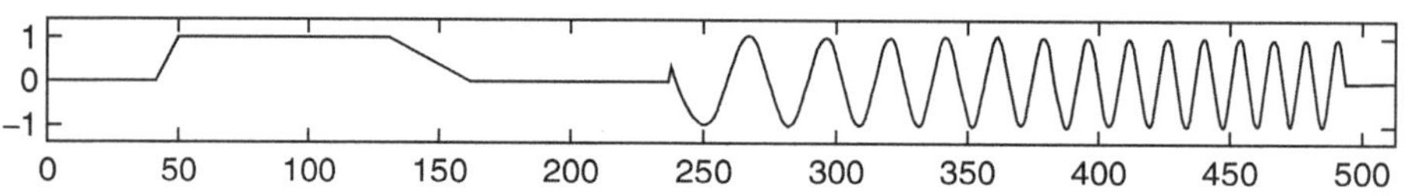

Fig. 5.5 Signal f to be analyzed [31]

From the previous example it follows that we can use the sombrero function to define a wavelet transform. We will use this wavelet to illustrate the flexibility of the wavelet transform in analyzing frequencies of a signal. For this, consider the signal whose graph is shown in Fig. 5.5.

This signal has high frequencies localized in the neighborhood of $t = 50$, and $t = 150$. From time $t = 280$, the signal has a chirp behavior: a continuum of increasing frequencies. In this region the signal is defined by the function

$$f(t) = \cos(t^3) .$$

We know already that the windowed Fourier transform is not adequate to analyze signals with this behavior. Figure 5.6 shows the graph of the signal and the graph of the wavelet transform for 5 distinct values of the scale s (the scale decreases from top to bottom).

Note that the frequencies associated with the sudden change of the signal at time $t = 50$ and time $t = 150$ are detected by the wavelet transform. Moreover, as the scale s decreases the high frequencies of the chirp signal $\cos(t^3)$ are also detected.

5.3 The Discrete Wavelet Transform

In the study of the windowed Fourier transform the time-frequency domain was discretized using a uniform lattice

$$\Delta_{(t_0,\omega_0)} = \{(mt_0, n\omega_0); m, n \in \mathbb{Z}\} ,$$

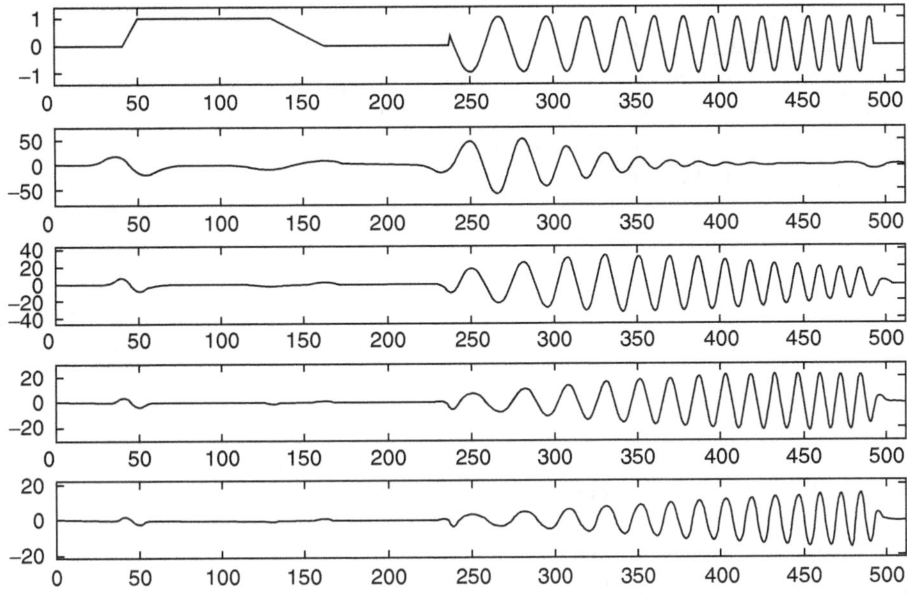

Fig. 5.6 The wavelet transform [31]

because of the constant width of the time-frequency atoms. The wavelet transform is defined on the time-scale domain. A natural question is:

Question 5.3. How to discretize the time-scale domain in such a way to obtain a discrete wavelet transform?

We know that the scaling operation acts in a multiplicative way, that is, composing two consecutive scalings is attained by multiplying each of the scale factors. Therefore the discretization of the scaling factor is simple: We fix an initial scale $s_0 > 1$, and we consider the discrete scales

$$s_m = s_0^m, \quad m \in \mathbb{Z} .$$

Positive values of m produce scales larger than 1, and negative values of m produce scales less than 1.

How to discretize the time? Initially we should observe that we must obtain a lattice in the time-scale domain in such a way that when we sample the wavelet transform $\tilde{f}(s,t)$ on this lattice, we are able to reconstruct the function f from the time-scale atoms $\tilde{f}_{m,n}$, with minimum redundancy. As the wavelet width changes with the scale, we must correlate the time with the scale discretization: As the scale increases the width of the wavelet also increases, therefore we can take samples further apart in the time domain. On the other hand, when the width of the wavelet decreases with a reduction of the scale, we must increase the frequency sampling.

To obtain the correct correlation between the scale and time discretization we observe that an important property of the wavelet transform is: *The wavelet transform is invariant by change of scales*. This statement means that if we make a change of scale in the function f and simultaneously change the scale of the underlying space by the same scaling factor, the wavelet transform does not change. More precisely, if we take

$$f_{s_0}(t) = s_0^{-1/2} f\left(\frac{t}{s_0}\right) ,$$

then

$$\tilde{f}_{s_0}(s_0 s, s_0 t) = \tilde{f}(s, t) .$$

Invariance by changing of scale constitutes an essential property of the wavelet transform. It is important that this property be preserved when we discretize the wavelet, so as to be also valid for the discrete wavelet transform. In order to achieve this goal, when we pass from the scale $s_m = s_0^m$ to the scale $s_{m+1} = s_0^{m+1}$, we must also increment the time by the scaling factor s_0. In this way, we can choose a time t_0 and take the length of the sampling time intervals as $\Delta t = s_0^m t_0$. Therefore, for each scale s_0^m the time discretization lattice is

$$t_{m,n} = n s_0^m t_0, \quad n \in \mathbb{Z} .$$

Finally, the discretization lattice in the time-scale domain is defined by

$$\Delta_{s_0,t_0} = \{(s_0^m, n s_0^m t_0) \; ; \; m, n \in \mathbb{Z}\} .$$

Example 11 (Dyadic Lattice). We will give a very important example of a wavelet discretization using $s_0 = 2$ (dyadic lattice). We have

$$\Delta_{2,t_0} = \{(2^m, n2^m t_0) \; ; \; m, n \in \mathbb{Z}\} .$$

The vertices of this lattice are shown in Fig. 5.7(a). This lattice is called *hyperbolic lattice* because it is a uniform lattice in hyperbolic geometry (only the points are part of the lattice).

To obtain a time-frequency lattice, we must observe that the frequency is the inverse of the scale. In this manner, for a given initial frequency ω_0 the lattice will be given by

$$\Delta_{2\omega_0,t_0} = \{(2^{-m}\omega_0, n2^{-m} t_0) \; ; \; m, n \in \mathbb{Z}\} .$$

The vertices of this lattice are shown in Fig. 5.7(b).

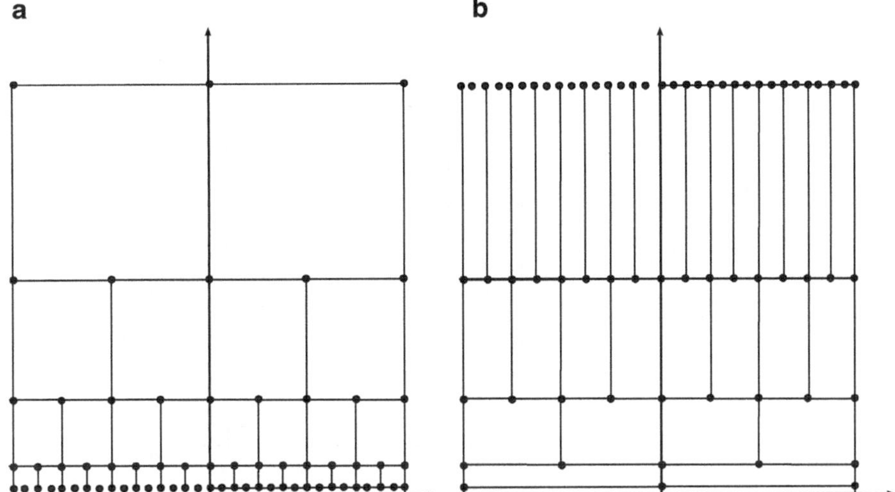

Fig. 5.7 **a** Time-scale lattice. **b** Time-frequency lattice

5.3.1 Function Representation

From the point of view of atomic decomposition the time-frequency atoms define a tiling of the time-frequency domain in rectangles as shown in Fig. 5.8.

The discretization of the wavelet transform $\tilde{f}(s,t) = \langle f, \psi_{s,t}(u)\rangle$ in the time-scale lattice is given by

$$\tilde{f}_{m,n} = \langle f, \psi_{m,n}(u)\rangle ,$$

where

$$\psi_{m,n}(u) = \psi_{s_0^m, nt_0 s_0^m}(u) \tag{5.14}$$

$$= s_0^{-m/2}\psi\left(\frac{u - nt_0 s_0^m}{s_0^m}\right) \tag{5.15}$$

$$= s_0^{-m/2}\psi(s_0^{-m}u - nt_0) . \tag{5.16}$$

In this context we can pose again the two questions which motivated the process of defining a discrete wavelet transform:

Question 5.4. Is the sequence $\langle f, \psi_{m,n}\rangle, m, n \in \mathbb{Z}$ an exact representation of the function f?

Question 5.5. Is it possible to reconstruct f from the family of wavelet time-frequency atoms $\psi_{m,n}$?

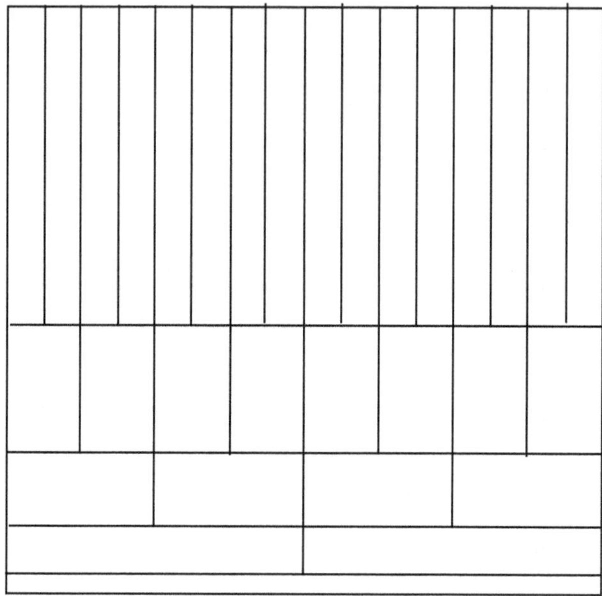

Fig. 5.8 Time-frequency decomposition using wavelets

A positive answer to these two questions would give us atomic decompositions of the function f using a family $\psi_{m,n}$ of discrete wavelets.

There are several directions we could take to answer the two questions above. Based on the representation theory discussed in Chap. 2, two natural questions in this direction are:

Question 5.6. Is it possible to define a lattice such that the corresponding family $\{\psi_{m,n}\}$ constitutes an orthonormal basis of $\mathbf{L}^2(\mathbb{R})$?

Question 5.7. Is it possible to define lattices for which the family $\{\psi_{m,n}\}$ is a frame?

If we have orthonormal basis of wavelets or a frame, we know from Chap. 2 that the answer to the two questions posed above is positive.

Chapter 3 of [20] brings a comprehensive discussion of frames of wavelets. The explicit construction of some wavelet frames is given. In the chapters to follow we will discuss the construction of different basis of wavelets.

Example 12 (Haar Basis). Consider the function

$$\psi(x) = \begin{cases} 1 & \text{if } x \in [0, 1/2) \\ -1 & \text{if } x \in [1/2, 1) \\ 0 & \text{if } x < 0 \text{ ou } x > 1. \end{cases}$$

The graph of f is shown in Fig. 5.9. This function satisfies the admissibility condition (5.8).

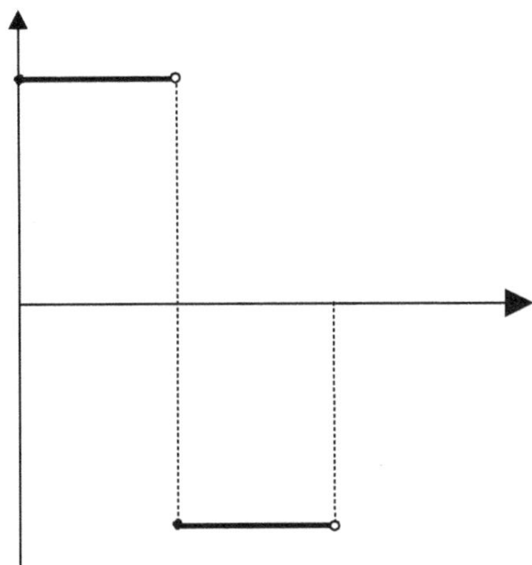

Fig. 5.9 Haar wavelet

It is possible to show that the set $\psi_{m,n}$, where

$$\psi_{m,n}(u) = 2^{-m/2}\psi(2^{-m}u - n), \quad m, n \in \mathbb{Z} ,$$

constitutes an orthonormal basis of $\mathbf{L}^2(\mathbb{R})$. Therefore we have an orthonormal basis of wavelets. A direct, and long, proof of this fact is found in [20], Sect. 1.3.3. The orthonormality of the set $\psi_{m,n}$ is easy to proof. The fact that the set generates the space $\mathbf{L}^2(\mathbb{R})$ is more complicated. This will follow as a consequence of the theory of multiresolution analysis that we will study in next chapter.

5.4 Comments and References

There are several possibilities of extending the wavelet transform to functions of several variables, i.e. $\mathbf{L}^2(\mathbb{R}^n)$. The interested reader should consult [20], p. 33, or [37].

The beautiful examples 9 and 10 of this chapter were taken from [31].

Chapter 6
Multiresolution Representation

In the introductory chapter we stated that two properties of wavelets were responsible for their applicability in the study of functions. One of these properties is the wavelet time-frequency atoms we studied in the previous chapter. The second property is the relationship of wavelets with multiresolution representation. This relationship will be exploited in two different ways:

- From one side it allows the use of wavelets to obtain multiresolution representations of functions.
- On the other hand, it will be used as a tool to construct wavelets.

6.1 The Concept of Scale

Our perception of the universe uses different scales: Each category of observations is done in a proper scale. This scale should be adequate to understand the different details we need. In a similar manner, when we need to represent an object, we try to use a scale where the important details can be captured in the representation.

A clear and well-known example of the use of scales occurs on maps. Using a small scale we can observe only macroscopic details of the mapped regions. By changing the scale we can observe or represent more details of the object being represented on the map.

Multiresolution representation is a mathematical model adequate to formalize the representation by scale in the physical universe. As we will see, this problem is intrinsically related to the wavelets.

The idea of scale is intrinsically related with the problem of point sampling of a signal. We call *sampling frequency* the number of samples in the unit of time. The length of the sample interval is called the *sampling period*. When we sample a signal using a frequency 2^j, we are fixing a scale to represent the signal: Details (frequencies) of the signal that are outside of the scale magnitude of the samples

J. Gomes, L. Velho, *From Fourier Analysis to Wavelets*,
IMPA Monographs 3, DOI 10.1007/978-3-319-22075-8_6

will be lost in the sampling process. On the other hand, it is clear that all of the details of the signal captured in a certain scale will also be well represented when we sample using a higher scale, 2^k, $k > m$.

These facts are well translated mathematically by the sampling theorem of Shannon-Whittaker that relates the sampling frequency with the frequencies present on the signal.

6.2 Scale Spaces

How to create a mathematical model to formalize the problem of scaling representation in the physical universe? The relation between sampling and scaling discussed above shows us the way. For a given integer number j, we create a subspace $V_j \subset \mathbf{L}^2(\mathbb{R})$, constituted by the functions in $\mathbf{L}^2(\mathbb{R})$ whose details are well represented in the scale 2^j. This means that these functions are well represented when sampled using a sampling frequency of 2^j.

The next step consists in creating a representation operator that is able to represent any function $f \in \mathbf{L}^2(\mathbb{R})$ in the scale 2^j. A simple and effective technique consists in using a representation by orthogonal projection. This is the Galerkin representation we discussed in Chap. 2. A simple and effective way to compute this representation is to obtain an orthonormal basis of V_j. But at this point we will demand more than that to make things easier: We will suppose that there exists a function $\phi \in \mathbf{L}^2(\mathbb{R})$ such that the family of functions

$$\phi_{j,k}(u) = 2^{-j/2}\phi(2^{-j}u - k), \quad j, k \in \mathbb{Z}\,, \tag{6.1}$$

is an orthonormal basis of V_j.

Notice that we are using here a process similar to the one we used when we introduced the wavelet transform: We define different scales of ϕ producing the continuous family

$$\phi_s(u) = \frac{1}{|s|^{1/2}}\phi\left(\frac{u}{s}\right) .$$

The width of ϕ and ϕ_s is related by

$$\text{width}(\phi) = s\ \text{width}(\phi_s) .$$

Thus, as the scale increases or decreases, the width of ϕ_s does the same. Equation (6.1) is obtained by discretizing the parameter s, taking $s = 2^j$, $j \in \mathbb{Z}$. Also, we have demanded that the translated family

$$\phi_{j,k} = \phi_{2^j}(u - k) = 2^{-j/2}\phi(2^{-j}u - k)$$

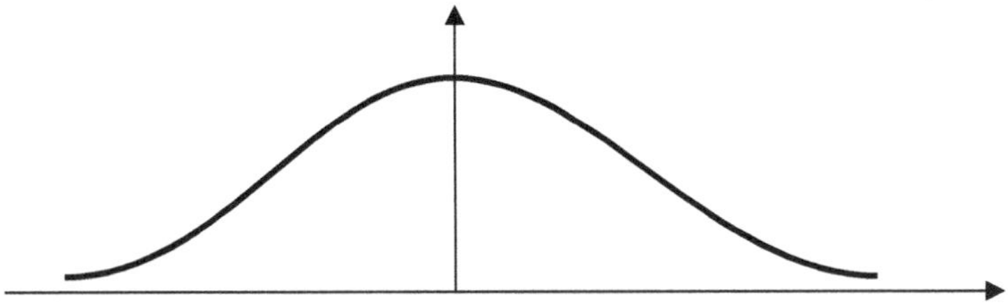

Fig. 6.1 Low-pass filter

is an orthonormal basis of V_j. Note that when j decreases, the width of $\phi_{j,k}$ also decreases, and the scale is refined. This means that more features of f are detected in its representation on the space V_j.

The representation of a function $f \in \mathbf{L}^2(\mathbb{R})$ by orthogonal projection in V_j is given by

$$\mathrm{Proj}_{V_j}(f) = \sum_k \langle f, \phi_{j,k} \rangle \phi_{j,k} .$$

We want the representation sequence $(\langle f, \phi_{j,k} \rangle)$ to contain samples of the function f in the scale 2^j. In order to attain this we know from Theorem 2 of Chap. 3 that the representation sequence $(\langle f, \phi_{j,k} \rangle)_{j,k \in \mathbb{Z}}$ is constituted by the samples of a filtered version of the signal f. More precisely,

$$\langle f, \phi_{j,k} \rangle = F(k) ,$$

where F is obtained from f by sampling with a filter of kernel $\phi_{j,k}$: $F = f * \phi_{j,k}$. In order that the elements of the representation sequence are close to the samples of f, the filter kernel $\phi_{j,k}$ must define a low-pass filter. This can be attained by demanding that $\hat{\phi}(0) = 1$, because $\hat{\phi}(\omega)$ approaches 0 when $\omega \to \pm\infty$. The graph of ϕ is depicted in Fig. 6.1. With this choice of ϕ, representing a function at scale 2^j amounts to sample averages of f over neighborhoods of width 2^j.

The space V_j is called *space of scale* 2^j, or simply *scale space*.

It is very important that we are able to change from a representation in a certain scale to a representation on another scale. For this we must answer the question: How are the different scale spaces related?

Since the details of the signal which appear on scale 2^j certainly must appear when we represent the signal using a smaller scale 2^{j-1}, we must have

$$V_j \subset V_{j-1} . \tag{6.2}$$

Given a function $f \in \mathbf{L}^2(\mathbb{R})$, a natural requirement is

$$f \in V_j \quad \text{if, and only if,} \quad f(2u) \in V_{j-1} . \tag{6.3}$$

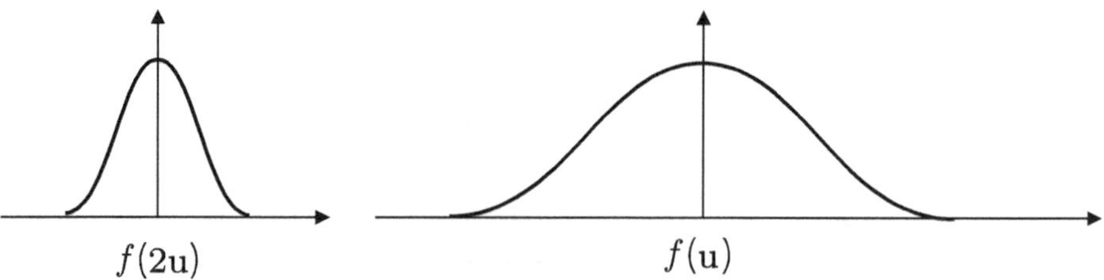

Fig. 6.2 Scaling of f by a scale factor of 2

In fact, the scaling of the variable of f by 2 reduces the width of f by the factor of $1/2$ (see Fig. 6.2). Therefore the details of f go to a finer scale.

Applying successively the condition in (6.3), we obtain

$$f \in V_j \quad \text{if, and only if,} \quad f(2^j u) \in V_0.$$

That is, all spaces are scaled version of the space V_0. In particular, from the fact that $\phi_{j,k}$ in Eq. (6.1) is an orthonormal basis of V_j, we conclude that

$$\phi_{0,k}(u) = \phi(u-k)$$

is an orthonormal basis of the scale space V_0.

The space $\mathbf{L}^2(\mathbb{R})$, our universe of the space of functions, contains all of the possible scales. This is reflected in the relation

$$\overline{\bigcup_{j\in\mathbb{Z}} V_j} = \mathbf{L}^2(\mathbb{R}) .$$

On the other hand, we have

$$\bigcap_{j\in\mathbb{Z}} V_j = \{0\} .$$

In effect, this expression says that the null function is the only function that can be well represented in every scale. In fact it should be observed that any constant function can be represented in any scale, nevertheless the only constant function that belongs to $\mathbf{L}^2(\mathbb{R})$ is the null function.

6.2.1 A Remark About Notation

It is important here to make a remark about the index notation we use for the scale spaces, because there is no uniformity in the literature. We use the notation of decreasing indices

$$\cdots V_1 \subset V_0 \subset V_{-1} \subset V_{-2} \cdots .$$

From the discussion above, this notation is coherent with the variation of the scale when we pass from one scale space to the other: As the indices decrease, the scale is refined, and the scale spaces get bigger.

If we use a notation with increasing indices

$$\cdots V_{-1} \subset V_0 \subset V_1 \cdots ,$$

which also appears in the literature, then the base $\phi_{j,k}$ of the scale space V_j should be constituted by the functions

$$\phi_{j,k}(x) = 2^{j/2}\phi(2^j x - k) .$$

This is rather confusing because it is not in accordance with the notation used when we discretized wavelets.

6.2.2 Multiresolution Representation

The scale spaces and their properties that we studied above define a multiresolution representation in $\mathbf{L}^2(\mathbb{R})$. We will resume them into a definition to facilitate future references:

Definition 1 (Multiresolution Representation). We define a multiresolution representation in $\mathbf{L}^2(\mathbb{R})$ as a sequence of closed subspaces V_j, $j \in \mathbb{Z}$, of $\mathbf{L}^2(\mathbb{R})$, satisfying the following properties:

(M1) $V_j \subset V_{j-1}$;
(M2) $f \in V_j$ if, and only if, $f(2u) \in V_{j-1}$.
(M3) $\bigcap_{j\in\mathbb{Z}} V_j = \{0\}$.
(M4) $\overline{\bigcup_{j\in\mathbb{Z}} V_j} = \mathbf{L}^2(\mathbb{R})$.
(M5) There exists a function $\phi \in V_0$ such that the set $\{\phi(u-k); k \in \mathbb{Z}\}$ is an orthonormal basis of V_0.

The function ϕ is called the *scaling function* of the multiresolution representation. Each of the spaces V_j is called *scale spaces*, or, more precisely, *space of scale* 2^j.

Example 13 (Haar Multiresolution Analysis). Consider the function

$$\phi(t) = \chi_{[0,1]} = \begin{cases} 0 & \text{if } x < 0 \text{ or } t \geq 1 \\ 1 & \text{if } x \in [0,1) . \end{cases}$$

It is easy to see that ϕ is a scale function of a multiresolution representation. In this case,

$$V_j = \{f \in \mathbf{L}^2(\mathbb{R}); f|[2^j k, 2^j(k+1)] = \text{constant}, k \in \mathbb{Z}\} .$$

That is, the projection of a function f on the scale space V_j is given by a function which is constant on the intervals $[2^j k, 2^j(k+1)]$. This is the Haar multiresolution representation.

We should notice that conditions (M1), ... (M5) that define a multiresolution representation are not independent. In fact it is possible to prove that condition (M3) follows from (M1), (M2), and (M5). Moreover, condition (M5) can be replaced by the weaker condition that the set $\{\phi(u-k)\}$ is a Riesz basis. Also, the reader might have noticed that we have not imposed that the scale function ϕ satisfies the condition $\hat{\phi}(0) = 1$ (as we know, this condition guarantees that ϕ is a low-pass filter). It can be proved that this low-pass filter condition follows from (M4). For a proof of all of these facts we suggest consulting [28] or [20]. We will return to this problem with more details in next chapter about construction of wavelets.

6.3 A Pause to Think

How to interpret geometrically the sequence of nested scale spaces in the definition of a multiresolution representation?

In general, visualizing subspaces of some space of functions is not an easy task. Nevertheless, in this case a very informative visualization of the nested sequence of scale space can be obtained in the frequency domain.

Indeed, the orthogonal projection of a function $f \in \mathbf{L}^2(\mathbb{R})$ in V_j is obtained using a filtering operation of f with the different kernels $\phi_{j,k}$, $k \in \mathbb{Z}$ which define low-pass filters. Indicating the cutting frequency of these filters by α_j (see Fig. 6.3), we conclude that each space V_j is constituted by functions whose frequencies are contained in the interval $[-\alpha_j, \alpha_j]$, $\alpha_j > 0$.

When we go from the space V_j to the space V_{j-1} we change from the scale 2^j to a finer scale 2^{j-1}. Therefore the frequency band increases to an interval $[-\alpha_{j-1}, \alpha_{j-1}]$. The graph of the spectrum of $\phi_{j-1,k}$ is the dotted curve in Fig. 6.3. The scale space V_{j-1} consists of the set of all the functions whose spectrum is contained in $[-\alpha_{j-1}, \alpha_{j+1}]$.

For each space V_j, with scale 2^j, we have the representation operator $R_j: L^2(\mathbb{R}) \to V_j$, given by the orthogonal projection over V_j

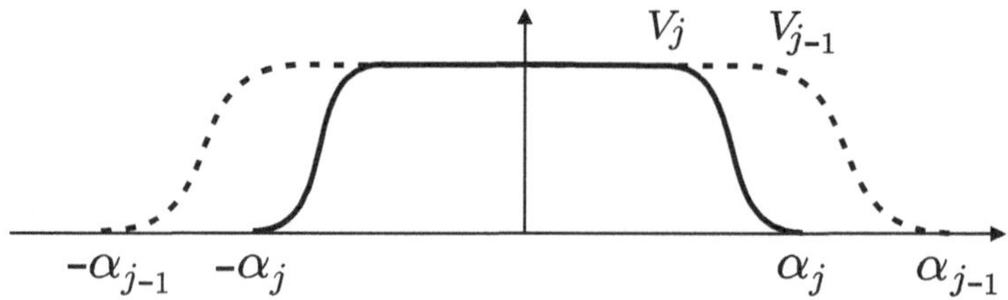

Fig. 6.3 Spectrum of the scaling function

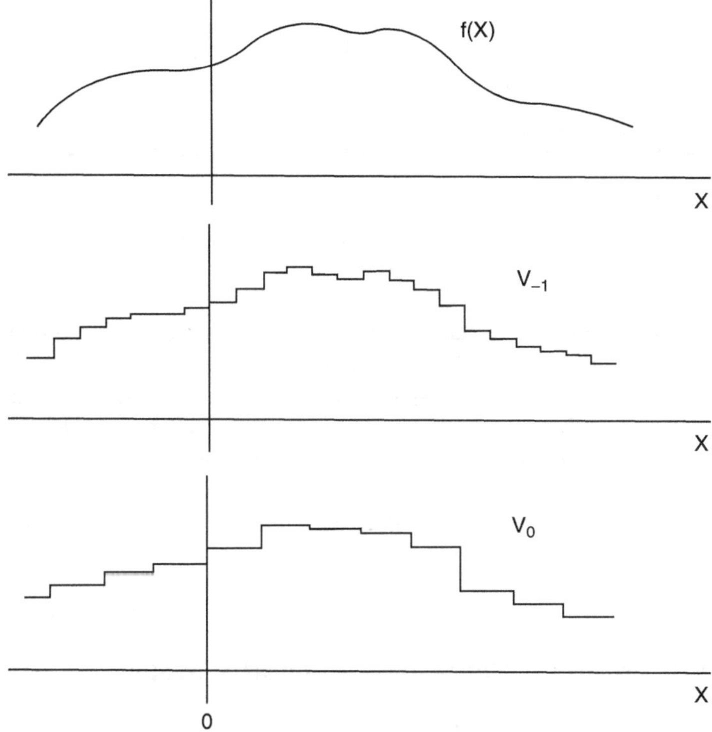

Fig. 6.4 Scale approximations of a function [20]

$$R_j(f) = \mathrm{Proj}_{V_j}(f) = \sum_k \langle f, \phi_{j,k} \rangle \phi_{j,k}.$$

From condition (M4) of the definition of a multiresolution representation, we have

$$\lim_{j \to \infty} R_j(f) = f, \tag{6.4}$$

that is, as the scale gets finer we get a better representation of the function f. This is illustrated in Fig. 6.4 (from [20]) we show a function f, and its representation on the spaces of scale V_0 and V_{-1} of the Haar multiresolution representation.

There is a different and very important way to interpret Eq. (6.4). Consider the graph representation of the space V_j in Fig. 6.5. We see that the space V_{j-1} is obtained from the space V_j by adding all of the functions from $\mathbf{L}^2(\mathbb{R})$ with frequencies in the band $[\alpha_j, \alpha_{j-1}]$ of the spectrum. We indicate this "detail space" by W_j. It follows immediately that W_j is orthogonal to V_j. Therefore we have

$$V_{j-1} = V_j \oplus W_j \ .$$

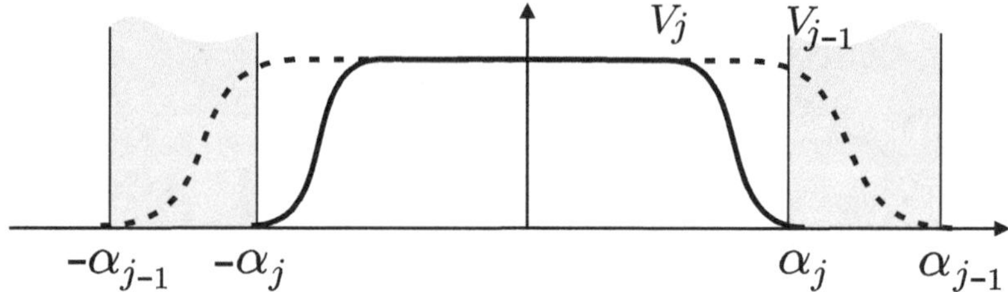

Fig. 6.5 Frequency band between V_j and V_{j-1}

The space W_j contains the details of the signal in the scale V_j. The above equation says that a function represented on a finer scale space V_{j-1} is obtained from the representation on a coarser scale space V_j, by adding details. These details can be obtained using a band-pass filtering, whose passband is exactly the interval $[\alpha_j, \alpha_{j-1}]$. We have seen that the wavelets constitute linear time-invariant band-pass filters. Therefore it seems natural that there might exist some relation between the detail spaces and the wavelets. We will discuss this relation "with details" in next section.

6.4 Multiresolution Representation and Wavelets

We have proved that given two consecutive scale spaces $V_j \subset V_{j-1}$, the orthogonal complement W_j of V_j in V_{j-1} could be obtained using a band-pass filter defined on $\mathbf{L}^2(\mathbb{R})$. In this section we will show that this complementary space is in fact generated by an orthonormal basis of wavelets.

For every $j \in \mathbb{Z}$, we define W_j as the orthogonal complement of V_j in V_{j-1}. We have

$$V_{j-1} = V_j \oplus W_j \, .$$

We remind that the best way to visualize the above equality is by observing the characterization of these spaces on the frequency domain (Fig. 6.5).

It is immediate to verify that W_j is orthogonal to W_k, if $j \neq k$. Therefore by fixing $J_0 \in \mathbb{Z}$, for every $j < J_0$ we have (see Fig. 6.6)

$$V_j = V_{J_0} \oplus \bigoplus_{k=0}^{J_0-j} W_{J_0-k} \, . \tag{6.5}$$

We should remark that because of the dyadic scales used in the discretization, the frequency bands do not have uniform length, they are represented in the figure using logarithmic scale.

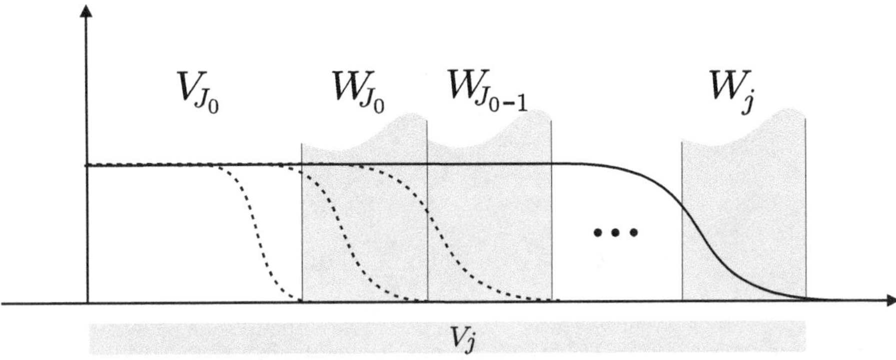

Fig. 6.6 Frequency bands between V_j and V_{J_o-j}

In sum, Eq. (6.5) says that the signals whose spectrum is in the frequency band of V_j, are the sum of the signals with frequency band in V_{J_0} with those signals whose frequency band is in $W_{J_0}, W_{J_0-1}, \ldots, W_j$. All of the subspaces involved in this sum are orthogonals. If $J_0, k \to \infty$, it follows from conditions (M3) and (M4) that define a multiresolution representation that

$$\mathbf{L}^2(\mathbb{R}) = \bigoplus_{j \in \mathbb{Z}} W_j ,$$

that is, we obtain a decomposition of $\mathbf{L}^2(\mathbb{R})$ as a sum of orthogonal subspaces.

We have seen that the projection of a function f in each subspace W_j could be obtained using a band-pass filter. In fact, this filtering process can be computed by projecting f on an orthogonal basis of wavelets. This fact is a consequence of the theorem below:

Theorem 5. *For each $j \in \mathbb{Z}$ there exists an orthonormal basis of wavelets $\{\psi_{j,k}, k \in \mathbb{Z}\}$ of the space W_j.*

We will sketch the proof of the theorem because it has a constructive nature which will provide us with a recipe to construct orthonormal basis of wavelets.

Basis of W_0 Initially we observe that the spaces W_j inherit the scaling properties of the scale spaces V_j. In particular,

$$f(u) \in W_j \quad \text{if, and only if,} \quad f(2^j u) \in W_0 \,. \tag{6.6}$$

For this reason, it suffices to show that there exists a wavelet $\psi \in W_0$ such that the set $\{\psi(u-k)\}$ is an orthonormal basis of W_0. In fact, in this case, it follows from (6.6) that the set

$$\{\psi_{j,k}(u) = 2^{-j/2}\psi(2^{-j}u - k)\}$$

is an orthonormal basis of W_j.

Low-pass filter and scaling function Since $\phi \in V_0 \subset V_{-1}$, and also $\phi_{-1,k}$ is an orthonormal basis of V_{-1}, we have

$$\phi = \sum_k h_k \phi_{-1,k} , \tag{6.7}$$

where

$$h_k = \langle \phi, \phi_{-1,k} \rangle, \quad \text{and} \quad \sum_{k \in \mathbb{Z}} ||h_k||^2 = 1 .$$

Substituting $\phi_{-1,k}(u) = \sqrt{2}\phi(2u - k)$ in (6.7) we obtain

$$\phi(x) = \sqrt{2} \sum_k h_k \phi(2x - k) . \tag{6.8}$$

Applying the Fourier transform to both sides of this equation, we have

$$\hat{\phi}(\xi) = m_o\left(\frac{\xi}{2}\right)\hat{\phi}\left(\frac{\xi}{2}\right) , \tag{6.9}$$

where

$$m_0(\xi) = \frac{1}{\sqrt{2}} \sum_k h_k e^{-ik\xi} .$$

Note in Eq. (6.9) that $\hat{\phi}\left(\frac{\xi}{2}\right)$ there exists a frequency band which has twice the size of the frequency band of $\phi(\xi)$. Therefore, it follows from (6.8) that the function m_0 is a low-pass filter. The function m_0 is called the *low-pass filter of the scaling function* ϕ. It is not difficult to see that m_0 is periodic with period 2π.

Characterization of W_0 Now we need to characterize the space W_0. Given $f \in W_0$, since $V_{-1} = V_0 \oplus W_0$, we conclude that $f \in V_{-1}$ and f is orthogonal to V_0. Therefore

$$f = \sum_n f_n \phi_{-1,n} , \tag{6.10}$$

where

$$f_n = \langle f, \phi_{-1,n} \rangle .$$

Computations similar to the ones we did to obtain the low-pass filter m_0 of the scaling function give us the equation

$$\hat{f}(\xi) = m_f\left(\frac{\xi}{2}\right)\hat{\phi}\left(\frac{\xi}{2}\right) , \tag{6.11}$$

where

$$m_f(\xi) = \frac{1}{\sqrt{2}} \sum_n f_n e^{-in\xi} .$$

After some computations, we can rewrite Eq. (6.11) in the form

$$\hat{f}(\xi) = e^{\frac{i\xi}{2}} \overline{m_0\left(\frac{\xi}{2} + \pi\right)} \nu(\xi) \hat{\phi}\left(\frac{\xi}{2}\right) , \tag{6.12}$$

where ν is a periodic function of period 2π.

Choosing the Wavelet Equation (6.12) characterizes the functions from W_0 using the Fourier transform, up to a periodic function ν. A natural choice is to define a wavelet $\psi \in W_0$ such that

$$\hat{\psi}(\xi) = e^{\frac{-i\xi}{2}} \overline{m_0\left(\frac{\xi}{2} + \pi\right)} \hat{\phi}\left(\frac{\xi}{2}\right) . \tag{6.13}$$

Taking this choice, from Eq. (6.12), it follows that

$$\hat{f}(\xi) = \left(\sum_k \nu_k e^{-ik\xi}\right) \hat{\psi}(\xi) ,$$

and applying the inverse Fourier transform, we have

$$f(x) = \sum_k \nu_k \psi(x - k) .$$

We need to show that defining ψ by the Eq. (6.13), $\psi_{0,k}$ is indeed an orthonormal basis of W_0. We will not give this proof here.

Details of the above proof can be found on [20] or [28].

6.5 A Pause... to See the Wavescape

If V_j is the scale space 2^j, we have $V_{j-1} = V_j \oplus W_j$. We know that W_j has an orthonormal basis of wavelets $\{\psi_{j,k}, k \in \mathbb{Z}\}$, therefore if R_j is the representation operator on the scale space V_j, we have, for all $f \in \mathbf{L}^2(\mathbb{R})$,

$$R_{j-1}(f) = R_j(f) + \sum_{k \in \mathbb{Z}} \langle f, \psi_{j,k} \rangle \psi_{j,k} . \tag{6.14}$$

The second term of the sum represents the orthogonal projection of the signal f on the space W_j and it will be denoted by $\text{Proj}_{W_j}(f)$. The terms of this representation sequence are obtained using the discrete wavelet transform.

We know that the wavelet transform is a band-pass filtering operation, and the scale spaces allow us to represent a function f in different resolutions. When we obtain a representation of f in a certain scale 2^j, we are losing details of the signal to respect with its representation in the scale 2^{j-1}. The lost details are computed by the orthogonal projection on the space W_j, that is,

$$\text{Proj}_{W_j}(f) = \sum_{k\in\mathbb{Z}} \langle f, \psi_{j,k}\rangle \psi_{j,k} , \tag{6.15}$$

which is a representation of the signal f in the basis of wavelets of the space W_j.

It is useful to interpret the decomposition $V_{j-1} = V_j \oplus W_j$ in the language of filters. The representation of a signal f in the scale V_j,

$$R_j(f) = \sum_{k\in\mathbb{Z}} \langle f, \phi_{j,k}\rangle \phi_{j,k} ,$$

is equivalent to filter the signal using the low-pass filter defined by the scaling function ϕ. The representation of the details of f in the space W_j, Eq. (6.15) is obtained by filtering f with the band-pass filter defined by the wavelet transform associated with ψ.

From the relation $V_{j-1} = V_j \oplus W_j$, we are able to write

$$\begin{aligned} R_{j-1}(f) &= R_j(f) + \text{Proj}_{W_j}(f) \\ R_{j-2}(f) &= R_{j-1}(f) + \text{Proj}_{W_{j-1}}(f) \\ &\vdots \end{aligned}$$

Note that each line of the equation above represents a low-pass filtering and a band-pass filtering of the signal. Iterating this equation for $R_{j-2}, \ldots, R_{j-J_0}$, summing up both sides and performing the proper cancellations, we obtain

$$R_{j-J_0}(f) = R_j(f) + \text{Proj}_{W_{j-1}}(f) + \cdots \text{Proj}_{W_{j-J_0}}(f) . \tag{6.16}$$

The projection $R_j(f)$ represents a version of low resolution (blurred version) of the signal, obtained using successive low-pass filtering with the filters $\phi_j, \phi_{j-1}, \ldots, \phi_{J_0-j}$. The terms $\text{Proj}_{W_{j-1}}(f)$, ..., $\text{Proj}_{W_{j-J_0}}(f)$ represent the details of the signal lost in each low-pass filtering. These details are obtained by filtering the signal using the wavelets $\psi_j, \psi_{j-1}, \ldots, \psi_{J_0-j}$. Equation (6.16) states that the original signal f can be reconstructed exactly from the low resolution signal, summing up the lost details.

6.6 Two-Scale Relation

In this section we will revisit some equations we obtained in the computations of this chapter in order to distinguish them for future references.

Consider a scaling function ϕ associated with some multiresolution representation. Then $\phi \in V_0 \subset V_{-1}$ and $\phi_{-1,n}$ is an orthonormal basis of V_{-1}. Therefore

$$\phi = \sum_{k\in\mathbb{Z}} h_k \phi_{-1,k} , \tag{6.17}$$

with $h_k = \langle \phi, \phi_{-1,k} \rangle$. This equation can be written in the form

$$\phi(x) = \sqrt{2} \sum_{k\in\mathbb{Z}} h_k \phi(2x - k) . \tag{6.18}$$

Similarly, given a wavelet ψ associated with a multiresolution representation $\psi \in V_0$, since $V_{-1} = V_0 \oplus W_0$, we have that $\psi \in V_{-1}$, and ψ is orthogonal to V_0, therefore

$$\psi = \sum_{k\in\mathbb{Z}} g_k \phi_{-1,k} , \tag{6.19}$$

or,

$$\psi(x) = \sqrt{2} \sum_{k\in\mathbb{Z}} g_k \phi(2x - k) . \tag{6.20}$$

Equations (6.17) and (6.19) (or equivalently (6.18) and (6.20)) are called *two-scale relations*, or *scaling relations* of the scaling function and the wavelet, respectively. In several important cases, the sum that defines the two-scale relations is finite:

$$\phi(x) = \sqrt{2} \sum_{k=0}^{N} g_k \phi(2x - k) .$$

It is not difficult to see that when this is the case, the support of the scaling function ϕ is contained in the interval $[0, N]$.

Also, note that if ϕ is a solution of the equation defined by the two-scale relation, then $\lambda\phi$, $\lambda \in \mathbb{R}$, is also a solution. In this way, to have uniqueness of the solution we must impose some kind of normalization (e.g., $\psi(0) = 1$).

A priori, it is possible to construct a multiresolution representation and the associated wavelet starting from an adequate choice of the function ϕ. This choice can be done using the two-scale relation (6.18). In a similar manner, the two-scale Eq. (6.20) can be used to obtain the associated wavelet.

6.7 Comments and References

The concept of multiresolution representation and its relation to wavelets was developed by S. Mallat [36]. In the literature it carries different names: *multiscale analysis* or *multiscale approximation*. We have opted for multiresolution representation because it fits better to the emphasis we have been given on function representation.

The material covered in this chapter can be found on [28]. Nevertheless the notation of the indices in the scale space differs from the one used here.

For an exposition of the topics in this chapter using the language of operators in function spaces the reader should consult [31]. The approach is algebraically very clear and clean, nevertheless a lot of geometric insight is lost.

Chapter 7
The Fast Wavelet Transform

The Fast Wavelet Transform (FWT) algorithm is the basic tool for computation with wavelets. The forward transform converts a signal representation from the time (spatial) domain to its representation in the wavelet basis. Conversely, the inverse transform reconstructs the signal from its wavelet representation back to the time (spatial) domain. These two operations need to be performed for analysis and synthesis of every signal that is processed in wavelet applications. For this reason, it is crucial that the Wavelet Transform can be implemented very efficiently.

In this chapter we will see that recursion constitutes the fundamental principle behind wavelet calculations. We will start with a revision of the multiresolution analysis to show how it naturally leads to recursion. Based on these concepts, we will derive the elementary recursive structures which form the building blocks of the fast wavelet transform. Finally, we will present the algorithms for the decomposition and reconstruction of discrete one-dimensional signals using compactly supported orthogonal wavelets.

7.1 Multiresolution Representation and Recursion

The efficient computation of the wavelet transform exploits the properties of a multiresolution analysis. In the previous chapters, we have seen that a multiresolution analysis is formed by a ladder of nested subspaces

$$\cdots V_1 \subset V_0 \subset V_{-1} \cdots$$

where all V_j are scaled versions of the central subspace V_0.

J. Gomes, L. Velho, *From Fourier Analysis to Wavelets*,
IMPA Monographs 3, DOI 10.1007/978-3-319-22075-8_7

From the above structure, we can define a collection of "difference" subspaces W_j, as the orthogonal complement of each V_j in V_{j-1}. That is,

$$V_j = V_{j+1} \oplus W_{j+1} .$$

As a consequence, we have a wavelet decomposition of $\mathbf{L}^2(\mathbb{R})$ into mutually orthogonal subspaces W_j

$$\mathbf{L}^2(\mathbb{R}) = \bigoplus_{j \in Z} W_j .$$

Therefore, any square integrable function $f \in \mathbf{L}^2(\mathbb{R})$ can be decomposed as the sum of its projection on the wavelet subspaces

$$f = \sum_{j \in \mathbb{Z}} \mathrm{Proj}_{W_j}(f)$$

where $\mathrm{Proj}_{W_j}(f)$ is the projection of f onto W_j.

From $V_j = V_{j+1} \oplus W_{j+1}$, it follows that any function $f_j \in V_j$ can be expressed as

$$f_j = \mathrm{Proj}_{V_{j+1}}(f) + \mathrm{Proj}_{W_{j+1}}(f) .$$

This fact gives us the main recursive relation to build a representation of a function using the wavelet decomposition.

If we denote the projections of f onto V_j and W_j, respectively, by $f_j = \mathrm{Proj}_{V_j}(f)$ and $o_j = \mathrm{Proj}_{W_j}(f)$, we can write

$$f_j = \underbrace{f_{j+1}}_{f_{j+2} + o_{j+2}} + o_{j+1} .$$

Applying this relation recursively we arrive at the wavelet representation

$$f_j = f_{j+N} + o_{j+N} + \cdots + o_{j+2} + o_{j+1}$$

where a function f_j in some V_j is decomposed into its projections on the wavelet spaces $W_{j+1} \ldots W_{j+N}$, and a residual given by its projection onto the scale space V_{j+N}. This recursive process can be illustrated by the diagram in Fig. 7.1.

$$\begin{array}{ccccccccc} f_j & \rightarrow & f_{j+1} & \rightarrow & f_{j+2} & \rightarrow & \cdots & \rightarrow & f_{j+N} \\ & \searrow & & \searrow & & \searrow & & \searrow & \\ & & o_{j+1} & & o_{g+1} & & \cdots & & o_{j+N} \end{array}$$

Fig. 7.1 Wavelet decomposition of a function f

$$\begin{array}{ccccccccc} f_{j+N} & \rightarrow & f_{j+N-1} & \rightarrow & \cdots & \rightarrow & f_{j+1} & \rightarrow & f_j \\ & \nearrow & & \nearrow & & \nearrow & & \nearrow & \\ o_{j+N} & & o_{j+N-1} & & \cdots & & o_{j+1} & & \end{array}$$

Fig. 7.2 Wavelet reconstruction process of a function f

We assumed above that the process starts with a function f_j which already belongs to some scale subspace V_j. This is not a restriction because we can take the initial j arbitrarily small (i.e., a fine scale). In practice, we work with functions that have some natural scale associated with them.

The wavelet decomposition gives an *analysis* of a function in terms of its projections onto the subspaces W_j. Note that, since by construction $W_j \perp W_l$ if $j \neq l$ and $V_j \perp W_j$, this decomposition of a function is unique once the spaces V_j and W_j are selected.

It is also desirable to reconstruct a function from its wavelet representation using a recursive process similar to the decomposition in Fig. 7.1. It turns out that, since $W_j \subset V_{j-1}$ and $V_j \subset V_{j-1}$, the original function can be obtained from the projections, and the wavelet reconstruction is essentially the reverse of the decomposition, as illustrated in Fig. 7.2.

The reconstruction gives a mechanism for the *synthesis* of functions from the wavelet representation.

To implement the wavelet decomposition and reconstruction we need to compute the projections onto the spaces V_j and W_j. We know that the set of functions $\{\phi_{j,n}; n \in \mathbb{Z}\}$ and $\{\psi_{j,n}; n \in \mathbb{Z}\}$, defined as

$$\phi_{j,n}(x) = 2^{-j/2}\phi(2^{-j}x - n) \tag{7.1}$$

$$\psi_{j,n}(x) = 2^{-j/2}\psi(2^{-j}x - n), \tag{7.2}$$

are, respectively, orthonormal basis of V_j and W_j. Therefore, the projection operators Proj_{V_j} and Proj_{W_j} are given by inner products with the elements of these bases

$$\text{Proj}_{V_j}(f) = \sum_n \langle f, \phi_{j,n} \rangle \phi_{j,n} = \sum_n \left(\int f(x)\overline{\phi_{j,n}(x)}dx \right) \phi_{j,n} \tag{7.3}$$

$$\text{Proj}_{W_j}(f) = \sum_n \langle f, \psi_{j,n} \rangle \psi_{j,n} = \sum_n \left(\int f(x)\overline{\psi_{j,n}(x)}dx \right) \psi_{j,n} \, . \tag{7.4}$$

The problem now is how to compute the projection operators Proj_{V_j} and Proj_{W_j} efficiently. In fact, we would like to avoid altogether computing the integrals explicitly. To find a solution we take advantage of the fact that the recursive decomposition/reconstruction processes require only projections between consecutive subspaces of the multiresolution ladder. For that purpose we will rely on the *two-scale* relations.

7.2 Two-Scale Relations and Inner Products

We have seen before that the interdependencies between two consecutive subspaces in a multiresolution analysis are formulated by the equations below, called *two-scale* relations

$$\phi(x) = \sum_k h_k \phi_{-1,k}(x) \tag{7.5}$$

$$\psi(x) = \sum_k g_k \phi_{-1,k}(x) \,. \tag{7.6}$$

Using these two relations, we can express the basis functions of the scale and wavelet spaces, V_j and W_j, at level j in terms of the basis functions of the subsequent scale space V_{j-1}, at finer level $j-1$. This is possible because, since $V_{j-1} = V_j \oplus W_j$, both $V_j \subset V_{j-1}$ and $W_j \subset V_{j-1}$.

Substituting (7.1) into (7.5), we have

$$\begin{aligned}
\phi_{j,k}(x) &= 2^{-j/2}\phi(2^{-j}x - k) \\
&= 2^{-j/2}\sum_n h_n\, 2^{1/2}\phi(2^{-j+1}x - 2k - n) \\
&= \sum_n h_n\, \phi_{j-1,2k+n}(x) \\
&= \sum_n h_{n-2k}\, \phi_{j-1,n}(x) \,.
\end{aligned} \tag{7.7}$$

Similarly, substituting (7.2) into (7.6), we have

$$\begin{aligned}
\psi_{j,k}(x) &= 2^{-j/2}\psi(2^{-j}x - k) \\
&= 2^{-j/2}\sum_n g_n\, 2^{1/2}\phi(2^{-j+1}x - 2k - n) \\
&= \sum_n g_{n-2k}\, \phi_{j-1,n}(x) \,.
\end{aligned} \tag{7.8}$$

Now, we need to find a way to use the sequences $(h_n)_{n\in\mathbb{Z}}$ and $(g_n)_{n\in\mathbb{Z}}$ to help us compute recursively the inner products $\langle f, \phi_{j,k}\rangle$, and $\langle f, \psi_{j,k}\rangle$. This can be easily done by inserting the expressions obtained for $\phi_{j,k}$ and $\psi_{j,k}$ into the inner products.

$$\langle f, \phi_{j,k}\rangle = \left\langle f, \sum_n \overline{h_{n-2k}}\phi_{j-1,n}\right\rangle = \sum_n \overline{h_{n-2k}}\langle f, \phi_{j-1,n}\rangle \tag{7.9}$$

$$\langle f, \psi_{j,k}\rangle = \left\langle f, \sum_n \overline{g_{n-2k}}\phi_{j-1,n}\right\rangle = \sum_n \overline{g_{n-2k}}\langle f, \phi_{j-1,n}\rangle \tag{7.10}$$

7.3 Wavelet Decomposition and Reconstruction

Using the two-scale relations, we showed how to relate the coefficients of the representation of a function in one scale 2^{j-1}, with the coefficients of its representation in the next coarse scale 2^j and with coefficients of its representation in the complementary wavelet space. It is remarkable that from the inner products of the function f with the basis of V_{j-1}, we are able to obtain the inner products of f with the basis of V_j and W_j, without computing explicitly the integrals! This is the crucial result for the development of the recursive wavelet decomposition and reconstruction method described in this section.

7.3.1 Decomposition

The wavelet decomposition process starts with the representation of a function f in the space V_0. There is no loss of generality here because, by changing the units, we can always take $j = 0$ as the label of the initial scale.

We are given the function $f = \text{Proj}_{V_0}(f)$, represented by the coefficients (c_k) of its representation sequence in the scale space V_0. That is

$$\text{Proj}_{V_0}(f) = \sum_k [\langle f, \phi_{0,k}\rangle \phi_{0,k}(x)] = \sum_k c_{0,k}\phi_{0,k} \;. \tag{7.11}$$

In case we only have uniform samples $f(k)$, $k \in \mathbb{Z}$ of the function, the coefficients (c_k) can be computed from the samples by a convolution operation. This fact is well explained in Sect. 3.7 of Chap. 3 (see Theorem 2).

The goal of the decomposition is to take the initial coefficient sequence $(c_k^0)_{k\in\mathbb{Z}}$, and transform it into the coefficients of the wavelet representation of the function. The process will be done by applying recursively the following decomposition rule

$$\text{Proj}_{V_j}(f) = \text{Proj}_{V_{j+1}}(f) + \text{Proj}_{W_{j+1}}(f) \;. \tag{7.12}$$

In this way, the process begins with $f^0 \in V_0 = V_1 \oplus W_1$, and in the first step, f^0 is decomposed into $f^1 + o^1$, where $f^1 = \text{Proj}_{V_1}(f)$ and $o^1 = \text{Proj}_{W_1}(f)$. The recursion acts on f^j, decomposing it into $f^{j+1} + o^{j+1}$, for $j = 0, \ldots N$. The components o^j are set apart. In the end we obtain the wavelet representation of f, consisting of the residual scale component f^N and the wavelet components $o^1, \ldots o^N$.

The core of the decomposition process splits the sequence (c_k^j) of scale coefficients associated with f^j into two sequences (c_k^{j+1}) and (d_k^{j+1}), of scale and wavelet coefficients associated, respectively, with f^{j+1} and o^{j+1}.

We can view this process as a basis transformation where we make the following basis change $(\phi_{j,k})_{k\in\mathbb{Z}} \to (\phi_{j+1,k}, \psi_{j+1,k})_{k\in\mathbb{Z}}$. Note that both sets form a basis of the space V^j. Equations (7.9) and (7.10) give the formulas to make the transformation on the coefficients of the bases:

$$c_k^{j+1} = \sum_n \overline{h_{n-2k}} c_n^j \tag{7.13}$$

$$d_k^{j+1} = \sum_n \overline{g_{n-2k}} c_n^j \tag{7.14}$$

with the notation $\overline{a} = (\overline{a_{-n}})_{n\in\mathbb{Z}}$.

Note that we are computing the coefficients (c_k^{j+1}) and (d_k^{j+1}) by discrete convolutions, respectively, with the sequences (h_n) and (g_n). Note also that we are retaining only the even coefficients for the next step of recursion (because of the factor $2k$ in the indices). This is a decimation operation.

In summary, if we start with a sequence (c_n^0), containing $n = 2^J$ coefficients, it will be decomposed into the sequences $(d_{n/2}^1)$, $(d_{n/4}^2)$, $\dots (d_{n/2^J}^J)$, and $(c_{n/2^J}^J)$. Note that the decomposition process outputs a wavelet representation with the *same number* of coefficients of the input representation.

Another important comment is that, up to now, we implicitly assumed doubly infinite coefficient sequences. In practice, we work with finite representations, and therefore it is necessary to deal with boundary conditions. This issue will be discussed in more detail later.

7.3.2 Reconstruction

The reconstruction process generates the coefficients of the scale representation from the coefficients of the wavelet representation. We would like to have an exact reconstruction, such that the output of the reconstruction is equal to the input of the decomposition. This is possible because we have just made an orthogonal basis transformation.

In order to bootstrap the recursive relations for the reconstruction process, we recall that one step of the decomposition takes a function representation f^{j-1} and splits into the components f^j and o^j.

$$\begin{aligned} f^{j-1}(x) &= f^j(x) + o^j(x) \\ &= \sum_k c_k^j \phi_{j,k}(x) + \sum_k d_k^j \psi_{j,k}(x) \,. \end{aligned} \tag{7.15}$$

We need to recover the coefficients (c_n^{j-1}) from (c^j) and (d^j)

$$c_n^{j-1} = \langle f^{j-1}, \phi_{j-1,n} \rangle \,. \tag{7.16}$$

Substituting (7.15) into (7.16), we obtain

$$c_n^{j-1} = \left\langle \sum_k c_k^j \phi_{j,k} + \sum_k d_k^j \psi_{j,k}, \quad \phi_{j-1,n} \right\rangle \tag{7.17}$$

$$= \sum_k c_k^j \langle \phi_{j,k}, \phi_{j-1,n} \rangle + \sum_k d_k^j \langle \psi_{j,k}, \phi_{j-1,n} \rangle \,. \tag{7.18}$$

Because both $\phi_0 \in V_{-1}$ and $\psi_0 \in V_{-1}$, they can be represented as a linear combination of the basis $\{\phi_{-1,n}; n \in \mathbb{Z}\}$. Therefore $\phi_0 = \sum_n \langle \phi_0, \phi_{-1,n} \rangle \phi_{-1,n}$ and $\psi_0 = \sum_n \langle \psi_0, \phi_{-1,n} \rangle \phi_{-1,n}$. Since this representation is unique, using the two-scale relations (7.5) and (7.6), we know that

$$h_n = \langle \phi_0, \phi_{-1,n} \rangle \tag{7.19}$$

$$g_n = \langle \psi_0, \phi_{-1,n} \rangle \,. \tag{7.20}$$

The above results provide a reconstruction formula for the coefficients c_n^{j-1} from the coefficient sequences of the decomposition at level j.

$$c_n^{j-1} = \sum_k h_{n-2k} c_k^j + \sum_k g_{n-2k} d_k^j$$

$$= \sum_k \left[h_{n-2k} c_k^j + g_{n-2k} d_k^j \right] \tag{7.21}$$

The reconstruction process builds the final representation (c_n^0), from bottom up. At each step, it combines the sequences (c_n^j) and (d_n^j) to recover the intermediate (c_n^{j-1}), from $j = J, \ldots, 1$.

7.4 The Fast Wavelet Transform Algorithm

The fast wavelet transform (FWT) algorithm is a straightforward implementation of the method described in the previous section. It consists of the recursive application of Eqs. (7.13) and (7.14) for the forward transform, and of Eq. (7.21) for the inverse transform.

In this section we present the pseudo-code, in C-like notation, of an implementation of the FWT algorithm. The code was structured for clarity and simple comprehension.

7.4.1 Forward Transform

The input of the algorithm is an array v, with 2^{m+1} elements, containing the coefficient sequence to be transformed, and the number of levels m. It uses the global arrays containing the two-scale sequences h and g. There are also global variables associated with these sequences: their number of elements hn and gn; and their offset values ho and go (i.e., the origins h_0 and g_0 of the sequences (h_n) and (g_n)). The main procedure wavelet_fwd_xform executes the iteration of the basic wavelet decomposition.

```
wavelet_fwd_xform(v, m, h, g)
{
        for (j = m; j >= 0; j--)
            wavelet_decomp(v, pow(2,j+1));
}
```

The procedure wavelet_decomp performs the decomposition for just one level, splitting the array v0 of size 2^{j+1} into two arrays v and w with sizes 2^j. The result is accumulated into the input array v, such that in the end of the decomposition the array v is partitioned into [vN | wN | ... | w2 | w1], with sizes, respectively, $1, 1, \ldots, 2^m, 2^{m-1}$.

```
wavelet_decomp(v, n)
{
        zero (w, 0, n);
        for (l = 0; l < n/2; l++) {
            i = (2*l + ho) % n;
            for (k = 0; k < hn; k++) {
                w[l] += v[i] * h[k];
                i = (i+1) % n;
            }
            i = (2*l + go) % n;
            m = l + n/2;
            for (k = 0; k < gn; k++) {
                w[m] += v[i] * g[k];
                i = (i+1) % n;
            }
        }
        copy (w, v, n/2);
}
```

The procedure uses a local array w that must have, at least, the same size of v. It calls two auxiliary procedures, zero that fills and array with zeros, and copy that copies one array to another.

7.4.2 Inverse Transform

The inverse transform takes as input an array containing the wavelet representation, in the format produced by `wavelet_fwd_xform`, and converts it into a scale representation.

The procedure `wavelet_inv_xform` executes the iteration of the basic reconstruction step.

```
wavelet_inv_xform(v, m)
{
    for (j = 0; j <= m; j++)
        wavelet_reconst(v, pow(2, j+1));
}
```

The procedure `wavelet_reconst` performs the reconstruction combining the components `vj` and `wj` of the input array to reconstruct `vj-1`. It replaces `[vj wj...]` with `[vj-1...]`. Note that the number of elements of `vj` and `wj` is $1/2$ of the number of elements of `vj-1`, therefore they use the same space in the array.

```
wavelet_reconst(w, n)
{
    zero(v, 0, n);
    for (k = 0; k < n; k++) {
        i = floor((k-ho)/2) % (n/2);
        m = (k - h.o) % 2;
        for (l = m; l < hn; l += 2) {
            v[k] += w[i] * h[l];
            i = (i-1) % (n/2);
        }
        i = floor ((k-go)/2) % (n/2);
        m = (k - go) % 2;
        for (l = m; l < gn; l += 2) {
             v[k] += w[i + n/2] * g[l];
             i = (i-1) % (n/2);
        }
    }
    copy(v, w, n);
}
```

7.4.3 Complexity Analysis of the Algorithm

The computational performance of the algorithm is very important. Let's determine what is the computational complexity of the fast wavelet transform.

The computation of each coefficient is a convolution operation with the two-scale sequences. Assuming that these sequences have n coefficients, then the convolution requires n multiplications and $n - 1$ additions.

In order to make the decomposition of a coefficient sequence at level j, from V_j into V_{j+1} and W_{j+1}, we have to compute 2^j new coefficients: 2^{j+1} for the two components f^{j+1} and o^{j+1}. Since each coefficient requires $2n - 1$ operations, we have a total of $2^j(2n - 1)$ operations for one-level transformation.

The full decomposition process is applied for $j \log_2(m)$ levels. Therefore, we have

$$\mathcal{O} = 2^j(2n - 1) + 2^{j+1}(2n - 1) + \cdots + 2(2n - 1)$$

factoring out $(2n - 1)$ and noting that $m = 2^j$, we obtain:

$$\mathcal{O}(m(2n - 1)[1 + 2^{-1} + 2^{-2} + \cdots + 2^{-j+1}])$$

$$\mathcal{O}\left(m(2n - 1)\frac{1 - 2^{-j}}{1 - 2^{-1}}\right)$$

$$\mathcal{O}(mn).$$

The above analysis leads us to the following conclusions:

- The complexity is linear with respect to the size of the input sequence;
- The size of the two-scale sequences has a direct relation with the algorithm complexity.

7.5 Boundary Conditions

Since in practice we work with finite sequences, it is necessary to take special care with the computation near the beginning and the end of the sequences (boundaries).

In order to compute the coefficients in the boundary regions, we have to perform a discrete convolution with the two-scale sequences, and therefore, we may need coefficients that lie beyond the boundaries of the sequence. Note that, for this reason, the boundary region is determined by size of the two-scale sequences. This situation is illustrated in Fig. 7.3.

There are some techniques to deal with boundary conditions:

- Extending the sequence with zeros (see Fig. 7.4(a));
- Periodization by translation of the sequence with $x(N + i) \equiv x(i)$ (Fig. 7.4(b));
- Periodization by reflection of the sequence with $x(N + i) \equiv x(N - i + 1)$ and $x(-i) \equiv x(i - 1)$ (Fig. 7.4(c));
- Use basis functions adapted to the interval (we are going to discuss this option later).

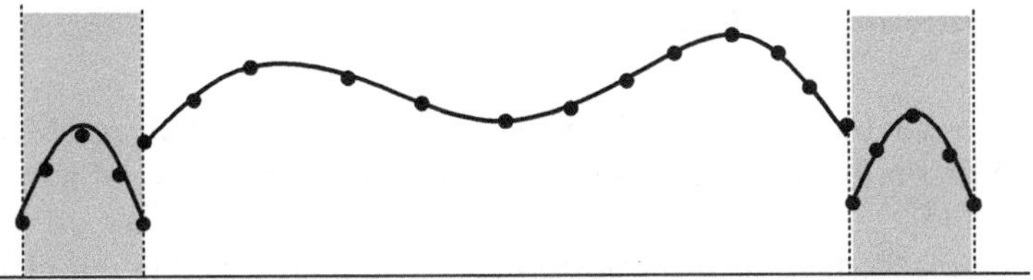

Fig. 7.3 Boundary regions for convolution between finite sequences

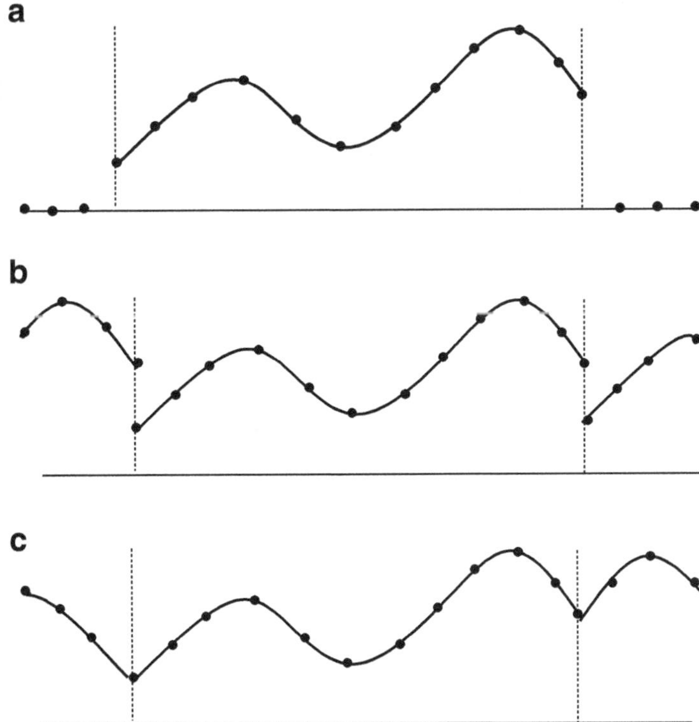

Fig. 7.4 Options for boundary computation. **a** Extending with zeros. **b** Periodization. **c** Reflection

In the implementation of the fast wavelet transform algorithm presented in Sect. 7.4, we deal with the boundary problem by a simple periodization of the sequence. This is accomplished using the coefficient with indices `i % m`.

7.6 Comments and References

The fast wavelet transform algorithm was introduced by Stephane Mallat [35]. One of the first references on the computational implementation of the algorithm appeared in [45].

The code for the fast wavelet transform algorithm presented in this chapter was based on the pseudo-code from [30]. This algorithm was implemented in [6].

The book [65] describes a complete system for computation with wavelets, including the fast wavelet transform.

Chapter 8
Filter Banks and Multiresolution

The goal of this chapter is to translate the theory of Multiresolution Representation to the language of Signal Processing.

Therefore, this chapter takes an important step in the change from the mathematical universe (continuous domain) to the representation universe (discrete domain), in the route to implementation.

The reader not familiar with signal processing will find basic concepts of linear systems and filters in Appendices A and B.

8.1 Two-Channel Filter Banks

We are going to study in greater detail a particular case of a type of filter bank that will be very important to understand the multiresolution representation in the discrete domain.

Consider a low-pass filter L and a high-pass filter H. We define an analysis filter bank S using those two filters together with downsampling operators as shown in Fig. 8.1.

Let's study the operations in the above system: the input signal (x_n) is processed by the filter L in order to obtain its low frequency components (y_{0n}), and also by the filter H to obtain its high frequency components (y_{1n}). After this first level of processing the filtered signals (y_{0n}) and (y_{1n}) constitute together a representation of the original signal (x_n), but with twice as much samples. In order to reduce the size of these two signals to the size of the original signal, we should discard terms. This can be achieved by performing a downsampling operation (see Appendix A). The output of the analysis bank, $S(x_n)$, is therefore a representation of the input signal in terms of its low and high frequency components, with the same size of the original signal. A natural question now is:

J. Gomes, L. Velho, *From Fourier Analysis to Wavelets*,
IMPA Monographs 3, DOI 10.1007/978-3-319-22075-8_8

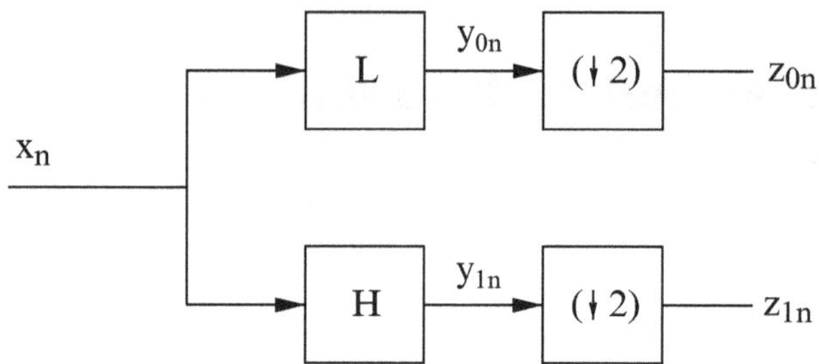

Fig. 8.1 Diagram of a Two-Channel Analysis Filter Bank

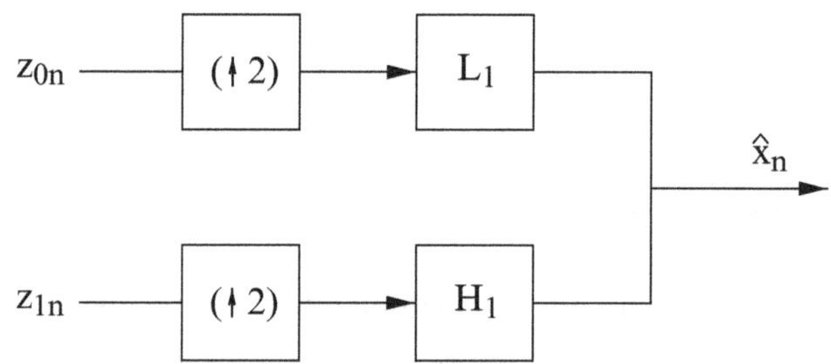

Fig. 8.2 Two-Channel Synthesis Filter Bank

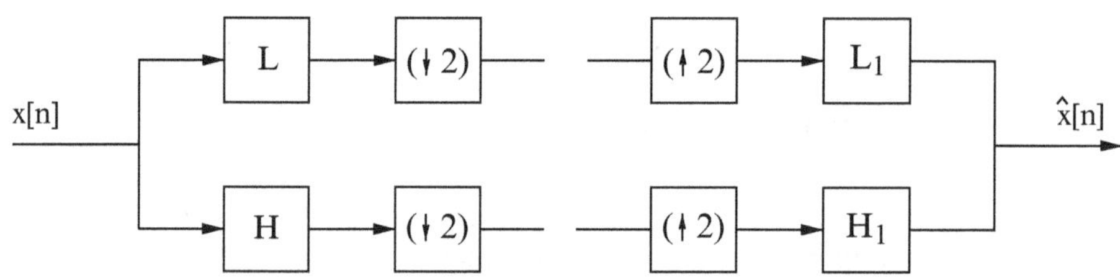

Fig. 8.3 Two-Channel Analysis and Synthesis Filter Bank

Question 8.1. Can we reconstruct the original signal (x_n) from the representation $S(x_n)$ produced by the analysis bank?

Because the downsampling operator $\downarrow 2$ is not invertible, the answer to this question on neither immediate nor obvious. An attempt of a solution is to define a synthesis filter bank $\tilde{S}$ as shown in Fig. 8.2.

More precisely, we apply an upsampling operator to the components of the two channels, low-pass and high-pass, and then use a pair of filters L_1 and H_1 such that these processed components when combined reconstruct the original signal.

The combination of these two filter banks S and $\tilde{S}$, forming a composite analysis and synthesis filter bank $\tilde{S} \circ S$, is shown in the diagram of Fig. 8.3.

The filter bank S is called *analysis bank* because it produces a representation of the signal in terms of its frequencies. The filter bank $\tilde{S}$ is called *synthesis bank* because it reconstructs the signal from its representation. When the output of the composite filter bank $\tilde{S} \circ S$ is the same as the input signal, that is

$$(\hat{x}_n) = \tilde{S}S(x_n) = (x_n) \ ,$$

we say that the filter bank has the property of *perfect reconstruction*. Usually, we have perfect reconstruction, but with a delay of the signal, which is not a problem because we can compensate that with a signal advance within the system.

8.1.1 Matrix Representation

Suppose that the filters L and H are defined as the convolution operators

$$L(x_n) = \sum_{k=0}^{3} a(k)x(n-k)$$

and

$$H(x_n) = \sum_{k=0}^{3} b(k)x(n-k) \ .$$

Then, the analysis bank is given by the matrix

$$S = \begin{pmatrix} \downarrow L \\ -- \\ \downarrow H \end{pmatrix} ,$$

or, using the results of applying the operator $\downarrow 2$ to a matrix,

$$S = \begin{pmatrix} a(3) & a(2) & a(1) & a(0) & & \\ & & a(3) & a(2) & a(1) & a(0) \\ b(3) & b(2) & b(1) & b(0) & & \\ & & b(3) & b(2) & b(1) & b(0) \end{pmatrix} .$$

It is clear that the perfect reconstruction property is related with the invertibility of the matrix S. In fact, if S is invertible, we can make the synthesis bank as $\tilde{S} = S^{-1}$, and we have exact reconstruction. An important particular case occurs when the linear system S is orthogonal, in this case the matrix of S is an orthogonal matrix. This system has the perfect reconstruction property and the synthesis filter, the inverse matrix $\tilde{S}$ is determined by the transpose of the analysis matrix S: $\tilde{S} = S^{\mathrm{T}}$.

Question 8.2. Why we have discussed two-channel filter banks with perfect reconstruction?

The answer to this question is easy: Given a multiresolution analysis, we know that the associated scaling function is a low-pass filter and the corresponding wavelet is a band-pass filter. We are going to show that, in the discrete domain, the multiresolution analysis defines a filter bank similar to the two-channel filter bank $\tilde{S} \circ S$, which has the property of perfect reconstruction. Furthermore, we will describe an algorithm that implements all the operations of the filter bank (analysis and synthesis) in linear time, proportional to the number of samples of the input signal (note that this is equivalent to the algorithm described in last chapter).

8.2 Filter Banks and Multiresolution Representation

In the mathematical universe (continuous domain) it is easy to convince ourselves that a multiresolution analysis defines a filter bank. This was already done earlier, but for the sake of reviewing the concepts and notation, we will repeat here.

If ϕ is the scaling function associated with the multiresolution analysis

$$\cdots \subset V_1 \subset V_0 \subset V_{-1} \subset \cdots$$

then $V_{j-1} = V_j \oplus W_j$, where W_j is the complementary (or wavelet) space. Given $f \in \mathbf{L}^2(\mathbb{R})$, we have that

$$\mathrm{Proj}_{V_{j-1}}(f) = \mathrm{Proj}_{V_j}(f) + \mathrm{Proj}_{W_j}(f) .$$

Also, $\mathrm{Proj}_{V_j}(f)$ represents the low-pass component of the function f, and $\mathrm{Proj}_{W_j}(f)$ represents the band-pass component of the function f both defined by the wavelet transform. Intuitively the low-pass component gives a representation of f in the scale 2^j, and the band-pass gives the details that are lost when f is represented in this scale.

Using the notation

$$L_j(f) = \mathrm{Proj}_{V_j}(f), \quad \text{and} \quad H_j(f) = \mathrm{Proj}_{W_j}(f) ,$$

we have

$$f = L_j(f) + H_j(f) .$$

Applying successively this decomposition to the component $L_j(f)$ of low frequencies of the signal, we arrive at

$$f = L_j^k(f) + H_k(f) + H_{k-1}(f) + \cdots + H_j(f) . \tag{8.1}$$

In summary, the signal f is decomposed into a low frequency component (i.e., in a scale that cannot capture details), $L_j^k(f)$, together with high frequency components, $H_n(f), n = k, k-1, \ldots, j$, which contain the details lost when going to a low resolution representation of f.

Equation (8.1) states that we can recover the function f exactly from the low resolution component by adding the high resolution components properly. Therefore, we have a (continuous) two-channel filter bank, with perfect reconstruction.

The decomposition operations of the analysis filter bank are illustrated by the diagram below

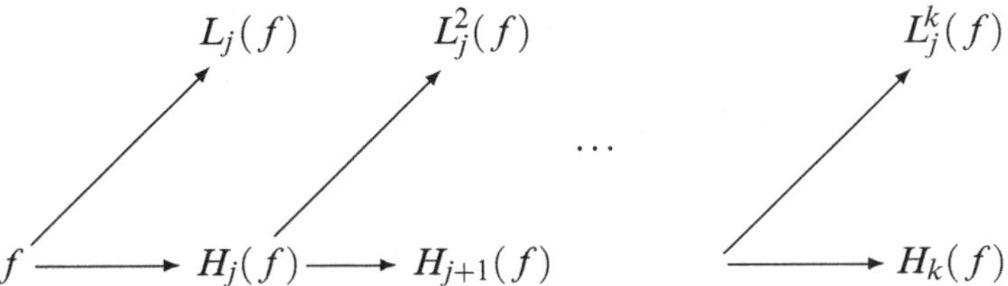

The reconstruction operations of the synthesis filter bank are illustrated by the diagram below

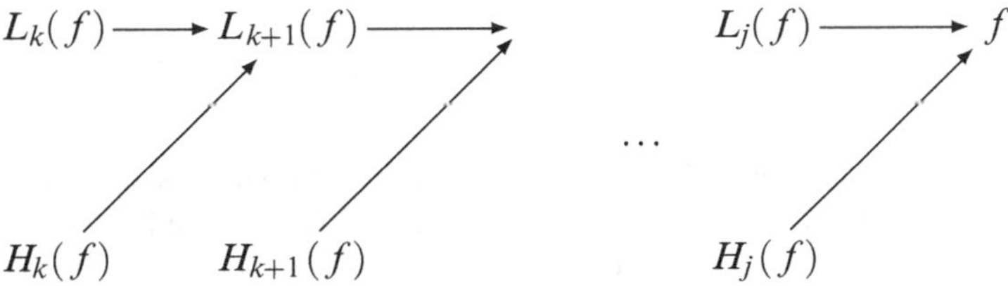

Our objective now is to study the above filter bank that corresponds to a multiresolution analysis, in the discrete domain.

8.3 Discrete Multiresolution Analysis

All ingredients for the discretization of a multiresolution analysis, and to revisit the associated filter bank in the discrete domain, are already of our knowledge.

If ψ is the wavelet associated with a multiresolution analysis, we have

$$\psi_{j,k}(x) = 2^{\frac{-j}{2}}\psi(2^{-j}x - k) . \tag{8.2}$$

We also have the two-scale relation that gives ψ as

$$\psi(x) = \sum_n g_n\phi_{-1,n}(x) = \sqrt{2}\sum_n g_n\phi(2x - n) ,$$

where ϕ is the corresponding scaling function, and $g_n = \langle \psi, \phi_{-1,n} \rangle$. Using this relation in (8.2), we have

$$\begin{aligned}\psi_{j,k}(x) &= 2^{\frac{-j}{2}} \sum_n g_n 2^{1/2} \phi(2^{-j+1}x - 2k - n) \\ &= \sum_n g_n \sum_n g_n 2^{\frac{-j+1}{2}} \phi(2^{-j+1}x - (2k+n)) \\ &= \sum_n g_n \phi_{j-1,2k+n}(x) .\end{aligned}$$

Making a change in the indices, we can rewrite the above equation in the form

$$\psi_{j,k}(x) = \sum_n g_{n-2k} \phi_{j-1,n} . \tag{8.3}$$

Then, it follows that

$$\langle f, \psi_{j,k}(x) \rangle = \sum_n \overline{g_{n-2k}} \langle f, \psi_{j-1,n} \rangle . \tag{8.4}$$

We can write the expression above using the filter operators so that the reader can more easily identify it with a filter bank. Indeed, it is immediate for the reader to verify that Eq. (8.4) above is equivalent to

$$\langle f, \psi_{j,k}(x) \rangle = (\downarrow 2)[(\langle f, \psi_{j-1,n} \rangle) * \overline{g_{-n}}] . \tag{8.5}$$

The two-scale relation for the function ϕ of the multiresolution analysis is given by

$$\phi = \sum_n h_n \phi_{-1,n} .$$

Computations similar to the ones we did from the two-scale relation for the wavelet give

$$\phi_{j,k}(x) = \sum_n h_{n-2k} \phi_{j-1,n} . \tag{8.6}$$

From that, it follows

$$\langle f, \phi_{j,k}(x) \rangle = \sum_n \overline{h_{n-2k}} \langle f, \phi_{j-1,n} \rangle . \tag{8.7}$$

As before, we can write this expression using the convolution and downsampling operators:

$$\langle f, \phi_{j,k}(x) \rangle = (\downarrow 2)[(\langle f, \phi_{j-1,n} \rangle) * \overline{h_{-n}}] . \tag{8.8}$$

8.3.1 Pause to Review

Eqs. (8.4) and (8.7) give the decomposition and reconstruction formulas of the filter bank associated with a multiresolution analysis. We are going to study these equations in more detail.

Without loss of generality, we can assume that the function f is defined initially in some scale as f^0. This scale is associated with some sampling frequency in which f can be represented by a sequence of samples. We suppose that $f^0 = \text{Proj}_{V_0}(f)$, that is f^0 is the discretization of f in the scale space V_0. Because $V_0 = V_1 + W_1$, we have that

$$f^0 = f^1 + o^1 .$$

Furthermore,

$$f^0 = \sum_n c_n^0 \phi_{0,n} ;$$

$$f^1 = \sum_n c_n^1 \phi_{1,n} ;$$

$$o^1 = \sum_n d_n^1 \phi_{1,n} .$$

From Eqs. (8.7) and (8.4), we have

$$c_k^1 = \sum_n h_{n-2k} c_n^0 ;$$

$$d_k^1 = \sum_n g_{n-2k} c_n^0 .$$

These equations allow us to obtain a representation of f in the finer scale V_{-1} from the initial representation sequence $(c_n^0)_{n\in\mathbb{Z}}$, in the scale V_0.

From a linear algebra point of view, we are just making a change of basis: From the basis $\{\phi_{0,n}\}_{n\in\mathbb{Z}}$ of the space V_0 to the basis $\{\phi_{1,n}, \psi_{1,n}\}_{n\in\mathbb{Z}}$.

Indicating by L the matrix of the operator in (8.5) and by H the matrix of the operator in (8.8), we can write

$$(c_n^1) = L(c_n^0) \quad \text{and} \quad (d_n^1) = H(c_n^0) .$$

Similarly,

$$f^1 \in V_1 = V_{-2} \oplus W_{-2}$$

and, therefore

$$f^1 = f^2 + o^2, \qquad f^2 \in V_{-2}, \quad o^2 \in W_{-2} .$$

Thus,

$$f^2 = \sum_n c_n^2 \phi_{2,n}, \quad o^2 = \sum_n d_n^2 \psi_{2,n} ,$$

with

$$(c_n^2) = L(c_n^1), \quad \text{and} \quad (d_n^2) = H(c_n^1) .$$

We have an analysis filter bank that is given by the diagram shown below

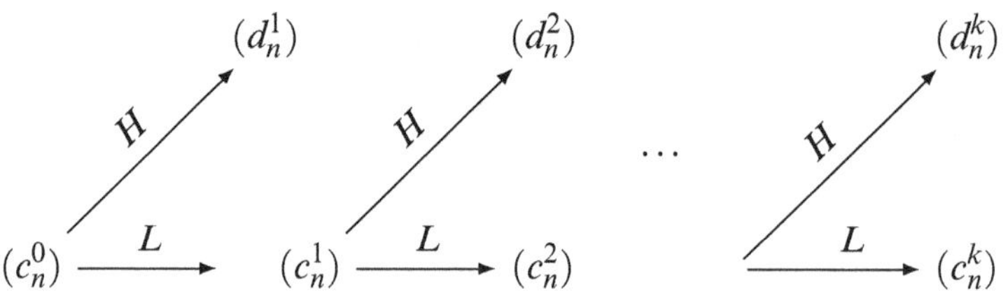

The above filter bank is a very particular case of a filter bank called *pyramid bank*. In fact, the pyramid structure is the best form to represent the type of filter bank in a diagram. For example, consider an initial signal given by a sequence of samples with 8 elements, that is,

$$f^0 = (c_0^0, c_1^0, c_2^0, \ldots, c_7^0).$$

The successive analyses of f^0 by the filter bank are represented in the inverted pyramid, shown in the diagram below:

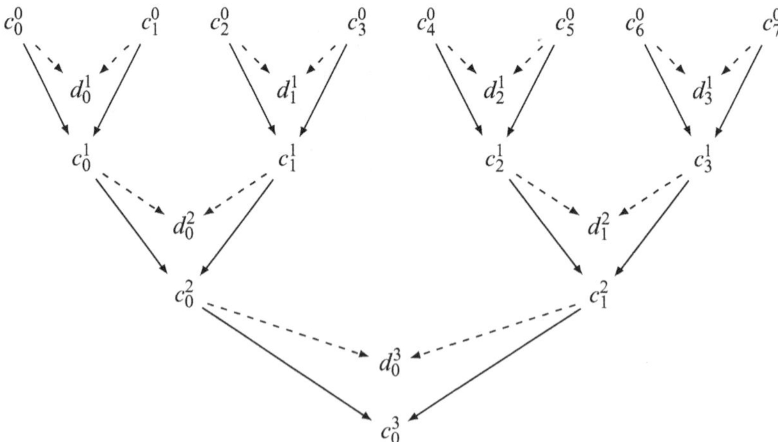

In the first level (base of the pyramid) we have the initial sequence of samples of f. In the second level we obtain the coefficients (c_n^1) of the lower resolution scale and the detail coefficients (d_n^1). We continue with the operation of the filter bank, going through each level of the pyramid until we reach a sequence with only one element (c_0^3) (which gives the average of the signal).

In matrix notation, the filter is represented by the structure

$$S = \begin{pmatrix} L_k & | & \\ H_k & | & \\ -- & - & - \\ & | & I \end{pmatrix} \cdots \begin{pmatrix} L_1 & | & \\ H_1 & | & \\ -- & - & - \\ & | & I \end{pmatrix} \begin{pmatrix} L \\ \\ - \\ H \end{pmatrix} . \tag{8.9}$$

The order of each block

$$\begin{pmatrix} L_k \\ -- \\ H_k \end{pmatrix}$$

with the low-pass filter L_k and high-pass filter H_k is half of the order of the preceding block

$$\begin{pmatrix} L_{k-1} \\ -- \\ H_{k-1} \end{pmatrix}$$

with filters L_{k-1} and H_{k-1}. Accordingly, the order of the block with the identity matrix doubles its size in each matrix. This product of matrices is the matrix representation of the filter bank of the multiresolution analysis. When this matrix is applied to the initial vector (c_n^0) it produces the complete multiresolution decomposition. The vector resulting from this operation $S((c_n^0))$ is

$$(c^{J_0}, d^{J_0}, d^{J_0-1}, \ldots, d^0) , \tag{8.10}$$

where

$$d^k = (d_0^k, d_1^k, \ldots, d_m^k), \quad k = J_0, J_0 - 1, \ldots, 0 .$$

Example 14 (Haar Multiresolution Analysis). The two-scale relations of the Haar multiresolution analysis are

$$\phi(t) = \phi(2t) + \phi(2t-1)$$
$$\psi(t) = \psi(2t) - \psi(2t-1) .$$

For a signal represented by four samples, the corresponding filter bank matrix in the first level has order 4 and is given by

$$\begin{pmatrix} L \\ G \end{pmatrix} = \begin{pmatrix} 1 & 1 & & \\ & & 1 & 1 \\ 1 & -1 & & \\ & & 1 & -1 \end{pmatrix} .$$

To make this matrix an orthogonal operator, we have to multiply its elements by $r = 1/\sqrt{2}$, obtaining the matrix

$$\begin{pmatrix} r & r & & \\ & & r & r \\ r & -r & & \\ & & r & -r \end{pmatrix} .$$

In the next level of scale the matrix is given by

$$\begin{pmatrix} r & r \\ r & -r \end{pmatrix} .$$

Therefore the matrix of the filter bank is

$$\begin{pmatrix} r & r & & \\ r & -r & & \\ & & 1 & \\ & & & 1 \end{pmatrix} \begin{pmatrix} r & r & & \\ & & r & r \\ r & -r & & \\ & & r & -r \end{pmatrix} .$$

8.4 Reconstruction Bank

Since the scaling function and wavelet are orthonormal bases, the operators involved into the analysis filter bank are also orthonormal. Therefore, the inverse operation for the synthesis filter bank is given by the adjoint matrix (i.e., the transpose of the conjugate). Indeed, the filter bank gives perfect reconstruction.

For completeness, we are going to derive the reconstruction expressions. We have

$$\begin{aligned} f^{j-1} &= f^j + d^j \\ &= \sum_k c_k^j \phi_{jk} + \sum_k d_k^j \psi_{j,k} . \end{aligned}$$

Therefore

$$
\begin{aligned}
c_n^{j-1} &= \langle f^{j-1}, \phi_{j-1,n} \rangle \\
&= \sum_k c_k^j \langle \phi_{j,k}, \phi_{j-1,n} \rangle + \sum_k d_k^j \langle \psi_{j,k}, \phi_{j-1,n} \rangle \\
&= \sum_k [h_{n-2k} c_k^j + g_{n-2k} d_k^j]
\end{aligned}
\tag{8.11}
$$

where we have used Eqs. (8.3) and (8.7) in the last line.

The Eq. (8.11) is the synthesis equation (reconstruction) of the filter bank associated with the multiresolution: it allows us to obtain the representation sequence (c_n^{j-1}) of the function f in a finer scale, from the sequences (c_n^j) and (d_n^j) in a lower resolution scale and its complement.

The reconstruction diagram is shown below

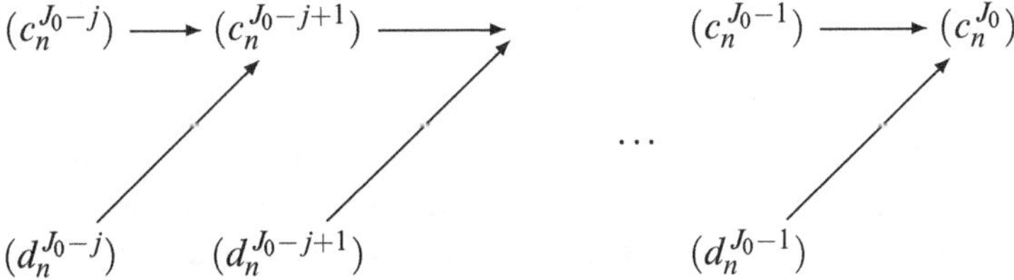

In summary, using a representation of the signal f at a low resolution C^{J_0-j} and the details corresponding to the information differences at intermediate resolutions $(d_n^{J_0-j})$, $(d_n^{J_0-j-1})$, ..., $(d_n^{J_0-1})$, we reconstruct exactly the representation c^{J_0} of the signal at the scale 2^{J_0}.

8.5 Computational Complexity

When we start with a function at some scale 2^j we have j levels in the pyramid, which correspond to j submatrices in the matrix that defines the filter bank of the multiresolution analysis. If the associated low-pass and high-pass filters have a kernel with T elements, we have T non-zero entries in each line of the matrix. Therefore, we have a product of TL non-zero elements in the computation of the first level of the filter bank, where $L = 2^j$ is the number of samples of the representation sequence of the input signal.

In the second level we have $TL/2$ products, in the third level $TL/4$ products, and so on. The number of products is reduced by one half in each level. Therefore, the total number of product to compute the analysis operation of the filter bank is given by

$$TL\left(1+\frac{1}{2}+\frac{1}{4}+\cdots+\frac{1}{2^{j-1}}\right) < 2TL\,.$$

We conclude that the computational complexity in the decomposition computation of a multiresolution analysis is linear with the length of the input signal.

In terms of matrices, the reconstruction is given by the conjugate transpose of the matrix S in (8.9), because the analysis operator is orthogonal. It follows trivially that the computational complexity of the reconstruction filter bank is the same as the decomposition bank: i.e. is also linear with the length of the input signal.

8.6 Comments and References

We have shown in this chapter that the multiresolution analysis defines a filter bank with perfect reconstruction. The filter bank has a pyramidal structure, and the decomposition and reconstruction operations can be computed in linear time relative to the length of the input signal.

The study of filter banks with perfect reconstruction started in the beginning of the 80s, and advanced independently of the studies in the area of wavelets, which intensified in the middle of the 80s (the Haar wavelet is known since the beginning of this century, but it was not associated with the context of wavelets).

We already mentioned in the previous chapter that the concept of multiresolution analysis was introduced by Stephane Mallat. In fact, Mallat also discovered the relation of filter banks and multiresolution analysis as we described in this chapter. He also developed the recursive algorithm to implement these operations, as we described in the previous chapter. This algorithm is known in the literature by various names: *Mallat algorithm*, *pyramid algorithm* or *fast wavelet transform*.

The multiresolution analysis is covered in various books [20, 28, 63] and [55]. This last reference [55] is one of the few which adopts a matrix notation.

Chapter 9
Constructing Wavelets

The proof of the existence of a wavelet associated with a multiresolution representation described in Chap. 6 has a constructive flavor. In this chapter we will go over details of this proof with the purpose of obtaining a recipe to construct wavelets.

9.1 Wavelets in the Frequency Domain

We start by analyzing the dilation equation, and looking at the bases functions ϕ and ψ in the frequency domain.

9.1.1 The Relations of $\hat{\phi}$ with m_0

Taking the Fourier transform of the two-scale equation

$$\phi(x) = \sqrt{2}\sum_n h_n \phi(2x - n) \tag{9.1}$$

gives

$$\hat{\phi}(\omega) = \frac{1}{\sqrt{2}}\sum_n h_n e^{-in\omega/2}\hat{\phi}(\omega/2) . \tag{9.2}$$

Equation (9.2) can be rewritten as

$$\hat{\phi}(\omega) = m_0(\omega/2)\hat{\phi}(\omega/2) \tag{9.3}$$

J. Gomes, L. Velho, *From Fourier Analysis to Wavelets*,
IMPA Monographs 3, DOI 10.1007/978-3-319-22075-8_9

with

$$m_0(\omega) = \frac{1}{\sqrt{2}} \sum_n h_n e^{-in\omega} . \tag{9.4}$$

The function m_0 is 2π-periodic, and $m_0 \in \mathbf{L}^2([0, 2\pi])$, because $\sum_{n\in\mathbb{Z}} |h_n|^2 < \infty$.

We also know that, by definition,

$$\int_{-\infty}^{\infty} \phi(x)dx = 1$$

hence, $\hat{\phi}(0) = 1$, and therefore

$$m_0(0) = 1 . \tag{9.5}$$

Applying Eq. (9.3) recursively for $w/2, w/4, \ldots$, we get $\hat{\phi}(\omega) = m_0(\omega/2)m_0(\omega/4)$ $\hat{\phi}(\omega/4)\ldots$, and arrive at the infinite product formula

$$\hat{\phi}(\omega) = \frac{1}{\sqrt{2\pi}} \prod_{j=1}^{\infty} m_0(2^{-j}\omega) . \tag{9.6}$$

A very important point is to show that this product converges to a function in $\mathbf{L}^2(\mathbb{R})$. Details of this proof can be found in [20].

We can see that $\hat{\phi}$ is completely characterized by m_0, as ϕ is completely characterized by the sequence (h_n). Note also that knowing m_0 gives us (h_n). This is the first important connection between wavelets in the spatial and frequency domains.

9.1.2 The Relations of $\hat{\psi}$ with m_1

Similarly, if we express the two-scale relation for the wavelet function ψ in the frequency domain,

$$\psi(x) = \sqrt{2} \sum_n g_n \psi(2x - n) \tag{9.7}$$

we get

$$\hat{\psi}(\omega) = \frac{1}{\sqrt{2}} \sum_n g_n e^{-in\omega/2} \hat{\phi}(\omega/2) \tag{9.8}$$

or

$$\hat{\psi}(\omega) = m_1(\omega/2)\hat{\phi}(\omega/2) \tag{9.9}$$

with

$$m_1(\omega) = \frac{1}{\sqrt{2}} \sum_n g_n e^{-in\omega} \tag{9.10}$$

where m_1 is also 2π-periodic.

Note that, $\hat{\psi}$ is defined in terms of $\hat{\phi}$ through m_1, in the same way ψ is defined in terms of ϕ through (g_n) in the spatial domain.

9.1.3 *Characterization of* m_0

In order to define the properties of m_0, we use the fact that $\phi(u-k)$, the integer translates of ϕ form an orthonormal basis of V_0. This imposes some restrictions on m_0.

$$\int_{-\infty}^{\infty} \phi(x)\overline{\phi(x-k)}dx = \int_{-\infty}^{\infty} |\hat{\phi}(\xi)|^2 e^{ik\xi} d\xi \;=\; \delta_{k,0} \tag{9.11}$$

$$= \int_0^{2\pi} e^{ik\xi} \sum_{l\in\mathbb{Z}} |\hat{\phi}(\xi + 2\pi l)|^2 d\xi \;=\; \delta_{k,0} \tag{9.12}$$

The above equation implies that

$$\sum_l |\hat{\phi}(\xi + 2\pi l)|^2 = \frac{1}{2\pi} \,. \tag{9.13}$$

Substituting Eq. (9.3) in (9.13), with $\omega = \xi/2$, we have

$$\sum_l |m_0(\omega + \pi l)|^2 |\hat{\phi}(\omega + \pi l)|^2 = \frac{1}{2\pi} \,. \tag{9.14}$$

We can split the sum into terms with even and odd l, and since m_0 is 2π-periodic we have

$$|m_0(\omega)|^2 \sum_l |\hat{\phi}(\omega+2l\pi)|^2 + |m_0(\omega+\pi)|^2 \sum_l |\hat{\phi}(\omega+(2l+1)\pi)|^2 = \frac{1}{2\pi} \tag{9.15}$$

substituting (9.13), and simplifying, we obtain

$$|m_0(\omega)|^2 + |m_0(\omega + \pi)|^2 = 1 \,. \tag{9.16}$$

This is the first important condition characterizing m_0, via orthonormality of ϕ.

If we put together Eq. (9.5) with (9.13), we conclude that

$$m_0(\pi) = 0\,. \tag{9.17}$$

This gives us a hint that m_0 is of the form (i.e., factorizes as)

$$m_0(\omega) = \left(\frac{1+e^{i\omega}}{2}\right)^m Q(\omega) \tag{9.18}$$

with $m \geq 1$, and where Q is a 2π-periodic function. (Observe that $e^{i\pi} = -1$. So, when $\omega = \pi$ the first term vanishes, and the product has to vanish.) We impose $Q(0) = 1$, to ensure that $m(0) = 1$, and also $Q(\pi) \neq 0$, so that the multiplicity of the root of m_0 at π is not increased by Q.

9.1.4 Characterization of m_1

Now, to link m_1 with m_0 we use the orthogonality between ϕ and ψ. More precisely, the constraint that $W_0 \perp V_0$ implies that $\psi \perp \phi_{0,k}$ and

$$\int_{-\infty}^{\infty} \hat{\psi}(\omega)\overline{\hat{\phi}(\omega)}e^{ik\omega}d\omega = 0 \tag{9.19}$$

or, in terms of the Fourier series

$$\int_0^{2\pi} e^{ik\omega} \sum_l \hat{\psi}(\omega + 2\pi l)\overline{\hat{\phi}(\omega + 2\pi l)}d\omega = 0 \tag{9.20}$$

hence

$$\sum_l \hat{\psi}(\omega + 2\pi l)\overline{\hat{\phi}(\omega + 2\pi l)} = 0 \tag{9.21}$$

for all $\omega \in \mathbb{R}$.

Substituting in the above equation the expressions (9.3) and (9.9) of $\hat{\phi}$ and $\hat{\psi}$ in terms of, respectively, m_0 and m_1, we obtain after regrouping the sums for odd and even l

$$m_1(\omega)\overline{m_0(\omega)} + m_1(\omega + \pi)\overline{m_0(\omega + \pi)} = 0\,. \tag{9.22}$$

This is the second important condition that characterizes m_0 and m_1.

We also know that $\overline{m_0(w)}$ and $\overline{m_0(w+\pi)}$ cannot be zero simultaneously because of (9.16), therefore m_1 can be written using m_0 and a function λ

$$m_1(\omega) = \lambda(\omega)\overline{m_0(\omega+\pi)} \tag{9.23}$$

such that λ satisfies

$$\lambda(\omega) + \lambda(\omega+\pi) = 0\,. \tag{9.24}$$

The simplest choice is $\lambda(\omega) = e^{iw}$, which gives m_1 satisfying the above equation

$$m_1(\omega) = e^{-i\omega}\overline{m_0(\omega+\pi)}\,. \tag{9.25}$$

Note that m_1 is defined in terms of m_0, as expected. This also gives $\hat{\psi}$ in terms of $\hat{\phi}$

$$\hat{\psi}(\omega) = e^{i\omega/2}\overline{m_0(\omega/2+\pi)}\hat{\phi}(\omega/2)\,. \tag{9.26}$$

From the above relations, we can construct an orthogonal wavelet from a scaling function ϕ, using (9.25) and choosing the coefficients $\{g_n\}$ as

$$g_n = (-1)^n h_{-n+1}, \tag{9.27}$$

that is,

$$\psi(x) = \sqrt{2}\sum_n (-1)^n h_{-n+1}\phi(2x-n)\,. \tag{9.28}$$

We conclude that, since m_1 is trivially defined from m_0, all we need to construct orthogonal scale and wavelet bases is to find a function m_0 satisfying (9.16) and (9.22), or, equivalently, find the coefficients (h_n) of the representation sequence of m_0.

Example 15 (Haar Wavelet). The scaling function of the Haar multiresolution representation is given by

$$\phi(x) = \begin{cases} 1 & \text{if } x \in [0,1) \\ 0 & \text{otherwise}\,. \end{cases}$$

From the two scaling relation

$$\phi(t) = \sqrt{2}\sum_k c_k\phi(2t-k),$$

we have

$$c_n = \sqrt{2}\int_{-\infty}^{+\infty} \phi(t)\phi(2t-n)dt\ .$$

It is easy to see that only $\phi(2t)$ and $\phi(2t-1)$ are not disjoint from $\phi(t)$, therefore only the coefficients of c_0 and c_1 are non-null (see Fig. 9.1). An easy computation shows that

$$c_0 = c_1 = \frac{\sqrt{2}}{2}\ .$$

Therefore the two-scale equation can be written as

$$\phi(x) = \phi(2x) + \phi(2x-1)\ .$$

This equation is illustrated in Fig. 9.1: In (a) we have the graph of ϕ and in (b) we have the sum of the functions $\phi(2x)$ and $\phi(2x-1)$.

The two-scale relation (9.28) for the wavelet in this case is given by

$$\psi(x) = \phi(2x) - \phi(2x-1),$$

which gives the Haar wavelet already introduced before (see Fig. 9.2).

Unfortunately the Haar example does not give a good view of the reality. The above reasoning is correct and promising but taking this trail to get to the wavelets involves a lot of work.

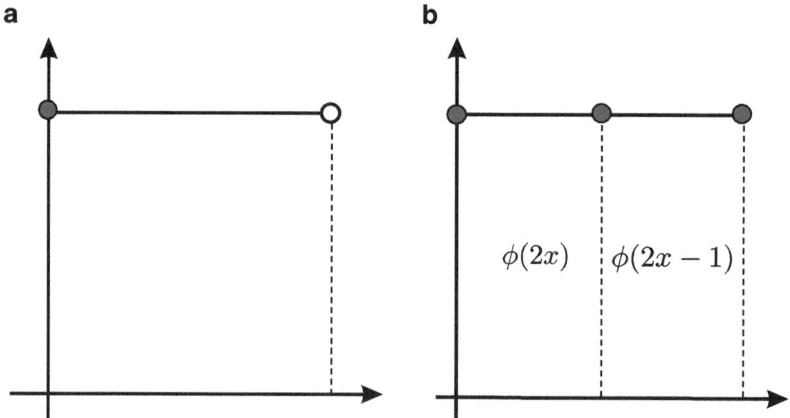

Fig. 9.1 Double scale relation of the Haar scaling function

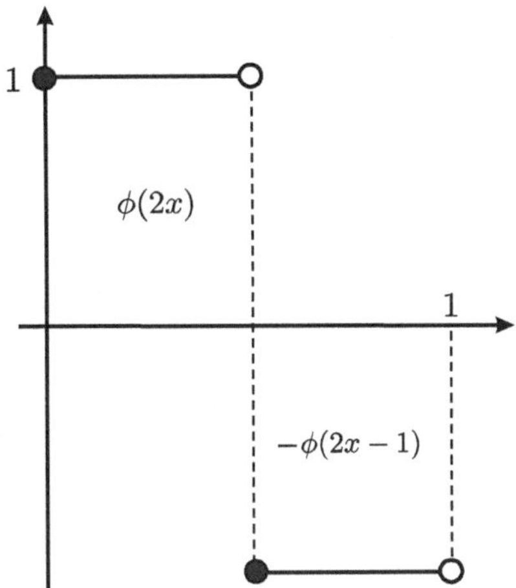

Fig. 9.2 Haar wavelet

9.2 Orthonormalization Method

Given a function $\phi(u)$ satisfying the two-scale relation (6.18), it may happen that the translated functions $\phi(x-n)$ do not constitute an orthonormal basis. We can overcome this fact.

In fact, we have stated before that the orthonormality condition (M5) in the definition of a multiresolution representation could be relaxed with the weaker condition that $\phi(x-n)$ is a Riesz basis. This fact is the essence of the orthonormalization process stated in the Theorem below:

Theorem 6. *If* $\{\phi(u-k\}$, $k \in \mathbb{Z}$, *is a Riesz basis of the space* V_0, *and we define a function* $\phi^{\#}$ *by*

$$\hat{\phi}^{\#}(\xi) = \frac{1}{\sqrt{2\pi}} \frac{\hat{\phi}(\xi)}{\sqrt{\sum_k |\hat{\phi}(\xi + 2\pi k)|^2}} , \tag{9.29}$$

then $\phi^{\#}(u-k)$, $k \in \mathbb{Z}$, *is an orthonormal basis of* V_0.

The proof of the theorem can be found in [20], p. 139.

In spite of using a weaker condition on (M5), the task of showing that a set of functions is a Riesz basis is not an easy one, in general. Moreover there exists the other conditions which must be satisfied in order to have a multiresolution representation. This problem can be solved using the result from

Theorem 7. *If* $\phi \in \mathbf{L}^2(\mathbb{R})$ *satisfies the two-scale relation*

$$\phi(x) = \sum_k c_k \phi(2x-k) ,$$

with

$$\sum_k |c_k|^2 < \infty ,$$

and

$$0 < \alpha < \sum_k |\hat{\phi}(\xi + 2\pi k)|^2 \leq \beta < \infty , \tag{9.30}$$

then ϕ defines a multiresolution representation.

9.3 A Recipe

In this section we will summarize the above results to construct a multiresolution representation, along with the associated wavelet.

Step 1 We start from a function ϕ which defines the kernel of a low-pass filter. A sufficient condition for this is that ϕ satisfies

$$\int_{\mathbb{R}} \phi(u)du \neq 0 ,$$

and also that ϕ and $\hat{\phi}$ have a good decay at infinity.

Step 2 We verify if the function ϕ of the previous step satisfies the two-scale relation (6.18) and the condition (9.30).

Step 3 If we have a function satisfying the two previous steps, but $\phi(u - k)$ is not an orthonormal basis, then we can obtain an orthonormal basis from ϕ, according to Theorem 6.

Step 4 The wavelet associated with the multiresolution representation can be computed using Eq. (6.13) from the previous chapter.

9.4 Piecewise Linear Multiresolution

The Haar multiresolution representation allows us to obtain successive approximations of a function using functions that are constant by parts. Now we will use the recipe from the previous section to obtain a multiresolution representation whose scale space approximates a function $f \in \mathbf{L}^2(\mathbb{R})$ using piecewise linear functions.

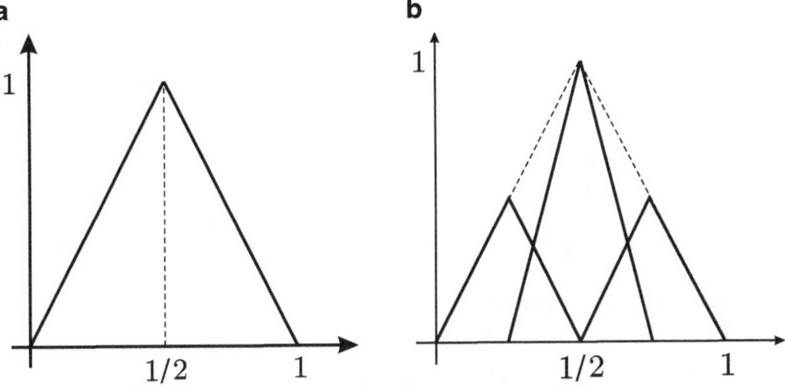

Fig. 9.3 Piecewise linear scaling function

We take as a candidate for the scaling function ϕ the function

$$\phi(x) = \begin{cases} 1 - |x| & \text{if } x \in [0, 1] \\ 0 & \text{if } x < 1 \text{ or } x > 1 \,. \end{cases}$$

The graph of ϕ is shown in Fig. 9.3(a).

Certainly the function ϕ satisfies all of the conditions in Step 1 of our recipe. Moreover, it is easy to see that ϕ also satisfies the two scaling relation

$$\phi(x) = \frac{1}{2}\phi(2x + 1) + \phi(2x) + \frac{1}{2}\phi(2x - 1)\,. \tag{9.31}$$

This relation is illustrated in Fig. 9.3(b).

The Fourier transform of ϕ is given by

$$\hat{\phi}(\omega) = \frac{1}{\sqrt{2\pi}}\left(\frac{\sin(\omega/2)}{\omega/2}\right)^2. \tag{9.32}$$

An immediate calculus shows that

$$2\pi \sum_{k\in\mathbb{Z}} |\hat{\phi}(\omega + 2\pi k|^2 = \frac{2}{3} + \frac{1}{3}\cos\omega = \frac{1}{3}\left(1 + \cos^2\left(\frac{w}{2}\right)\right)\,. \tag{9.33}$$

Therefore the condition (9.30) is satisfied. From Theorem 7, ϕ defines a multiresolution representation.

Nevertheless, it is easy to verify that the family $\phi(u - k), k \in \mathbb{Z}$, is not orthonormal. Therefore we must apply the orthonormalization process (Theorem 6) to obtain a scaling function that defines a multiresolution orthonormal basis.

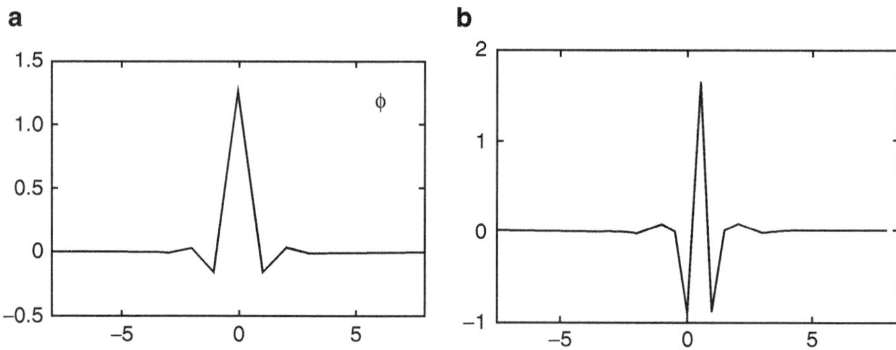

Fig. 9.4 **a** Scaling function of the piecewise linear multiresolution. **b** Wavelet of the piecewise linear multiresolution

Substituting (9.32) and (9.33) in (9.29), we have

$$\hat{\phi}^{\#}(\xi) = \sqrt{\frac{3}{2\pi}} \frac{4\sin^2\left(\frac{\xi}{2}\right)}{\xi^2\left(1 + 2\cos^2\left(\frac{\xi}{2}\right)\right)^{1/2}},$$

that is the Fourier transform of the scaling function we are looking for. The graph of $\hat{\phi}^{\#}$ is shown in Fig. 9.4(a). The graph of the associated wavelet $\psi(x)$ is shown in Fig. 9.4(b).

9.5 Shannon Multiresolution Analysis

From classical Fourier analysis we know the low-pass ideal filter $\phi\colon \mathbb{R} \to \mathbb{R}$, whose transfer function is given by

$$\hat{\phi}(\omega) = \chi_{[-\pi,\pi]} = \begin{cases} 1 & \text{if } x \in [-\pi, \pi] \\ 0 & \text{if } x < -\pi \text{ or } x > \pi \,. \end{cases}$$

The impulse response of the filter is given by the function

$$\phi(x) = \operatorname{sinc}(x) = \frac{\sin(\pi x)}{\pi x}\,.$$

The graph of this function is shown in Fig. 9.5.

It is natural to expect that the function ϕ defines a multiresolution representation. This is in fact true, but we will not give the details here. It is called *Shannon multiresolution representation*. Shannon multiresolution representation divides the frequency domain into bands according to the illustration shown in Fig. 9.6.

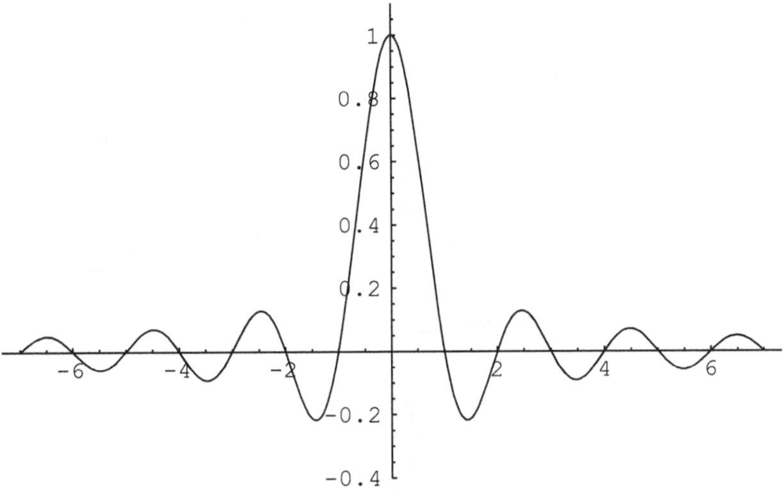

Fig. 9.5 Impulse response of the ideal reconstruction filter

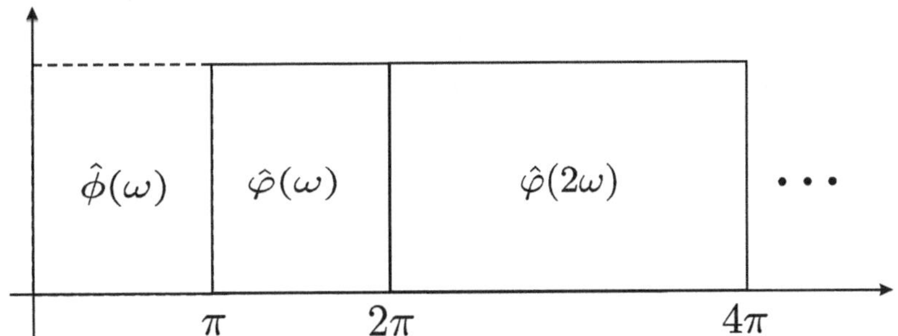

Fig. 9.6 Band decomposition using Shannon multiresolution representation

The wavelet ψ associated with Shannon multiresolution representation is the ideal band-pass filter, defined by

$$\hat{\psi} = \chi_I, \quad \text{where} \quad I = [-2\pi, -\pi) \cup (\pi, 2\pi] \, .$$

The graph of this filter is shown in Fig. 9.7.

This wavelet is called *Shannon wavelet*. It has an analytic expression given by

$$\psi(x) = -2\frac{\sin(2\pi x) + \cos(2\pi x)}{\pi(2x + 1)} \, .$$

The graph is shown in Fig. 9.8.

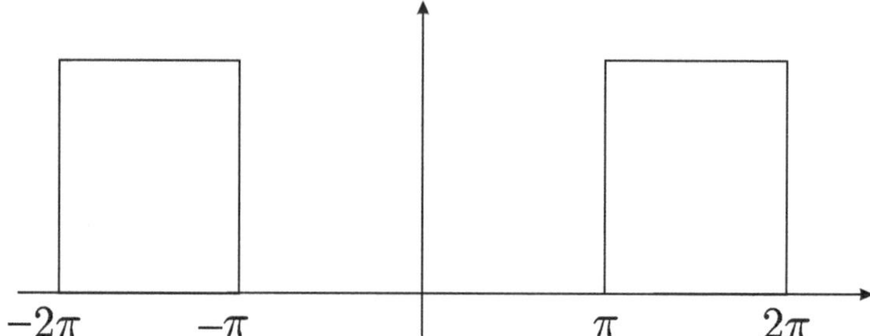

Fig. 9.7 Ideal band-pass filter

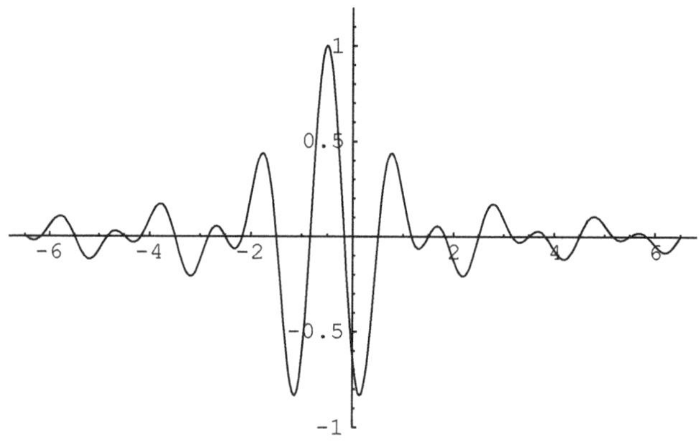

Fig. 9.8 Shannon wavelet

9.6 Where to Go Now?

Up to this point we have seen how to obtain atomic decompositions using wavelets, we have studied the role the wavelets in the multiresolution representation, we have shown the existence of orthonormal basis of wavelets and we have described a method to construct such basis using multiresolution representation.

Certainly, in applications we need to have at our disposal a great abundance of basis of wavelets. Therefore the natural path to follow should point in the direction of devising methods to construct basis or frames of wavelets.

In order to amplify the range of possible applications, we should impose additional conditions to the wavelets we construct. Among many conditions we could mention:

- regularity;
- symmetry;
- compact support;
- orthogonality with polynomials of degree $\leq n$.

The construction of wavelets satisfying some of the above properties, as well as a discussion about the importance of these properties in different types of applications is of fundamental importance. In Chap. 10, we will present a framework for generating wavelets using a filter design methodology.

9.7 Comments and References

In this chapter we have described how wavelets can be constructed. Devising robust techniques to construct wavelets is a very important research topic. The reader should realize that each application demands bases of wavelets with different properties. We will cover this in the chapters to follow.

The Haar wavelet and the wavelet associated with the piecewise linear multiresolution representation are part of a family of wavelets called *spline wavelets*. This family is related to a multiresolution representation that approximates a function $f \in \mathbf{L}^2(\mathbb{R})$ using piecewise polynomials of degree $\leq n$. These types of wavelets are studied in [13].

Chapter 10
Wavelet Design

We have seen in the previous chapters that orthogonal wavelets can be defined from a multiresolution analysis. This framework provides the basic relations to construct wavelets and to compute the fast wavelet transform.

As we discussed in Chap. 9, the scaling and wavelet functions, ϕ and ψ, are completely characterized by the two-scale sequences (h_n) and (g_n), respectively. Therefore, a direct method to generate new wavelets consists in finding ϕ and ψ that satisfy the two-scale relations, and whose integer translates form orthonormal bases of V_0 and W_0.

This simple approach can be used to produce examples of wavelets, but it presents two main difficulties: first, it does not give a systematic way to find the functions ϕ and ψ; second, it relies on an orthonormalization step which produces infinitely supported bases.

It is clear that we need a more effective and flexible method to design wavelets. In this chapter, we present an approach to generate wavelets based on a frequency domain analysis. We will exploit the fact that the two-scale sequences are the coefficients of a two-channel discrete filter bank. Consequently, we can use the filter design methodology for creating wavelets.

10.1 Synthesizing Wavelets from Filters

In this section we review the restrictions on the filter function m_0, with the purpose of discovering an expression for it that can be computed.

J. Gomes, L. Velho, *From Fourier Analysis to Wavelets*,
IMPA Monographs 3, DOI 10.1007/978-3-319-22075-8_10

10.1.1 Conjugate Mirror Filters

The functions m_0 and m_1 can be interpreted as the discrete Fourier transform of a pair of discrete filters $H = (h_n)$ and $G = (g_n)$, respectively, as we discussed in the previous chapter. The function m_0 is a low-pass filter for the interval $[0, \pi/2]$, and the function m_1 is a band-pass filter for the interval $[\pi/2, \pi]$.

From these observations, and from the definition of ϕ and ψ in the frequency domain, we conclude that the main part of the energy of $\hat{\phi}$ and $\hat{\psi}$ is concentrated, respectively, in the intervals $[0, \pi]$ and $[\pi, 2\pi]$.

The fact that $m_0(0) = m_1(\pi)$ and $m_1(0) = m_0(\pi)$ together with the relation $m_1(\omega)\overline{m_0(\omega)} + m_1(\omega + \pi)\overline{m_0(\omega + \pi)} = 0$ makes H and G a pair of filters that are complementary. These filters are called *conjugate mirror filters*, because their frequency responses are mirror images with respect to the middle frequency $\pi/2$ (also known as the quadrature frequency).

The above means that the wavelet transform essentially decomposes the frequency space into dyadic blocks $[2^j\pi, 2^{j+1}\pi]$ with $j \in \mathbb{Z}$.

Note that it is possible to construct m_0 and m_1 that are quadrature mirror filters, but do not correspond to any functions ϕ and ψ in $\mathbf{L}^2(\mathbb{R})$ as defined by a multiresolution analysis. In order to guarantee that infinite product formula (9.6) will converge to the Fourier transform of a valid ϕ, the function m_0 must satisfy $|m_0(\omega)|^2 + |m_0(\omega + \pi)|^2 = 1$, the orthogonality condition for $\phi(\cdot - k)$. In addition, we have to impose extra conditions on m_0 to ensure that $\sum_l |\hat{\phi}(\omega + 2\pi l|^2 = 1/2\pi$. (See [20] for the technical details.)

10.1.2 Conditions for m_0

First, let's revisit the two main conditions, (9.16) and (9.22), that the function m_0 must satisfy to generate orthogonal scaling functions and wavelets. They can be summarized as:

$$\begin{aligned} &\text{Condition 1:} \quad m_0(\omega)\overline{m_0(\omega)} + m_0(\omega + \pi)\overline{m_0(\omega + \pi)} = 1 \\ &\text{Condition 2:} \quad m_1(\omega)\overline{m_0(\omega)} + m_1(\omega + \pi)\overline{m_0(\omega + \pi)} = 0 \end{aligned}$$

where, in the first equation we expanded $|m_0(\omega)|^2 = m_0(\omega)\overline{m_0(\omega)}$.

Notice that these two conditions can be written in matrix form as

$$\begin{pmatrix} m_0(\omega) & m_0(\omega + \pi) \\ m_1(\omega) & m_1(\omega + \pi) \end{pmatrix} \overline{\begin{pmatrix} m_0(\omega) \\ m_0(\omega + \pi) \end{pmatrix}} = \begin{pmatrix} 1 \\ 0 \end{pmatrix}$$

or $Mx = y$, where M is called the *modulation matrix*, in the filter bank theory.

$$M = \begin{pmatrix} m_0(\omega) & m_0(\omega + \pi) \\ m_1(\omega) & m_1(\omega + \pi) \end{pmatrix} .$$

We will return to this matrix later, in the more general context of biorthogonal wavelets and perfect reconstruction filters.

For orthogonal wavelets, we were able to transform Condition 2, into

$$m_1(\omega) = e^{-i\omega}\overline{m_0(\omega + \pi)} ,$$

using the quadrature mirror filter construction (see Eq. (9.25) of previous chapter). Therefore, we only need to determine $m_0(\omega)$ such that Condition 1 is satisfied.

The goal now is to go from Condition 1 to a formula for m_0.

10.1.3 Strategy for Computing m_0

What we would really like at this point is to obtain a closed form expression for m_0. Unfortunately, this would not be possible. There are no simple formulas for the coefficients (h_n) of m_0. But, we will be able to compute (h_n) numerically.

Instead of working directly with m_0, we will consider $|m_0|^2$. We define the function $P(\omega)$ as the product $m_0(\omega)\overline{m_0(\omega)}$

$$P(\omega) = \left(\sum_{l=0}^{N} h_l e^{il\omega}\right)\left(\sum_{n=0}^{N} h_n e^{-in\omega}\right)$$

$$= \sum_{k=-N}^{N} a_k e^{-ik\omega} . \tag{10.1}$$

Our strategy is to decompose the problem of finding m_0 satisfying Condition 1 into two simpler sub-problems:

1. Find a function $P(\omega)$ that satisfy $P(\omega) + P(\omega + \pi) = 1$;
2. From P obtain m_0, factoring $P(\omega)$ into $m_0(\omega)\overline{m_0(\omega)}$.

It turns out that it is better to work with P than with m_0, because using P, the analysis of Condition 1 is much simpler. This strategy makes possible to find an explicit expression for $P(\omega)$, and therefore, the formulas to compute its coefficients (a_k). From that, we can obtain m_0 through the spectral factorization of P (i.e., taking a "square root").

We will see, in the next two chapters, how to derive a closed form expression for $P(\omega)$, respectively, in the context of orthogonal scaling functions, where $P(\omega) = m_0(\omega)\overline{m_0(\omega)}$, and in the more general context of biorthogonal scaling functions, where $P(\omega) = m_0(\omega)\overline{\widetilde{m_0}(\omega)}$ (i.e., the functions m_0 and $\widetilde{m_0}$ can be different).

In the rest of this chapter, we will assume that P is known, and we will concentrate on the details of how to get m_0 from P.

10.1.4 Analysis of P

First, to simplify the exposition, we will make a change of notation. We express $re^{i\omega} = z$, with $r = |z|$, to convert from the ω notation to the z notation. Let $P(z)$ restricted to the unit circle $r = 1$ denote $P(\omega)$. So, Eq. (10.1) becomes

$$P(z) = \sum_{k=-N}^{N} a_k z^k \ . \tag{10.2}$$

There are many things we can say about P, even without knowing the specific expression of P.

- P is a Laurent polynomial of degree N, with both positive and negative exponents. This is apparent in Eq. (10.2).
- $P(z) \leq 0$ is real and non-negative, because $P(z) = |m_0(z)|^2$ (in the orthogonal case).
- P has real coefficients a_k, $k = -N, \ldots, N$, because m_0 has real coefficients.
- P is symmetric, with $a_k = a_{-k}$, because $m_0(z)$ multiplies its conjugate $\overline{m_0(z)}$.

Taking into consideration also Condition 1, we can say more about P. Using the fact that $e^{iw+\pi/2} = -e^{iw}$, we write Condition 1 in the z notation as:

$$P(z) + P(-z) = 1 \ . \tag{10.3}$$

The terms with odd coefficients cancel in Eq. (10.3) (and there is nothing we can say about them). But, we can conclude that, except for the constant term, all other even coefficients of P have to be *zero* to make (10.3) true.

$$a_{2m} = \begin{cases} 1/2 & \text{if } m = 0 \\ 0 & \text{if } m \neq 0 \end{cases} \tag{10.4}$$

In the filter theory, the function P which satisfy (10.3) is called a *halfband filter*. When it comes from $P(z) = |m_0(z)|^2$, it is the autocorrelation filter of m_0. It gives the power spectral response, and m_0 is the spectral factor. Figure 10.1 shows a graph of P.

10.1.5 Factorization of P

The function $P(z)$ is a Laurent polynomial of degree N. Since it is symmetric there are $N + 1$ independent coefficients in P. In the next two chapters, we will derive a formula to compute these coefficients. Assuming that the coefficients of P are known, the remaining problem is to factor P to obtain m_0.

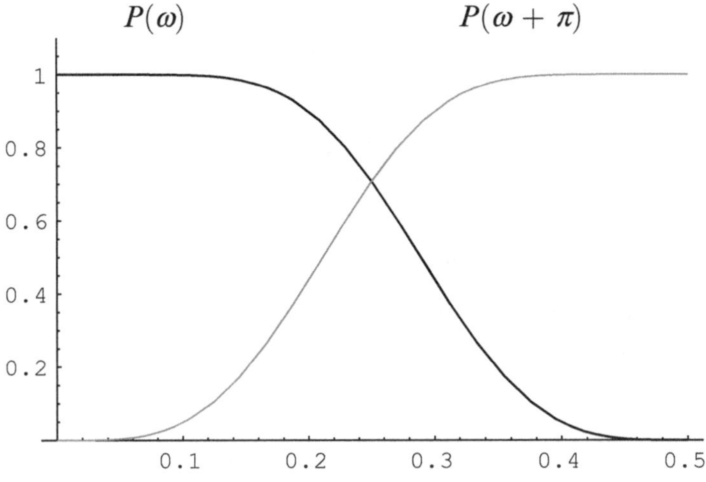

Fig. 10.1 Halfband Filter

A regular polynomial of degree d can be represented in two forms:

$$B(z) = \sum_{n=0}^{d} b_n z^n = \prod_{i=1}^{d} (z - z_i)$$

where b_n are the coefficients of B, and z_i are the roots of B.

The factorization of the polynomial P can be done by finding the zeros (roots) of P. There are several methods for this purpose. One of the most effective of these methods consists in computing the eigenvalues the companion matrix. The roots of a regular polynomial B of degree d are the eigenvalues of the $d \times d$ companion matrix

$$\mathbf{B} = \begin{pmatrix} \frac{-b_{d-1}}{b_d} & \frac{-b_{d-2}}{b_d} & \cdots & \frac{-b_1}{b_d} & \frac{-b_0}{b_d} \\ 1 & 0 & \cdots & 0 & 0 \\ 0 & 1 & \cdots & 0 & 0 \\ \vdots & & \ddots & & \vdots \\ 0 & 0 & \cdots & 1 & 0 \end{pmatrix} \tag{10.5}$$

where $B(z) = \det(\mathbf{B} - z\mathbf{I})$ is the characteristic polynomial of $\mathbf{B}$.

In order to apply the factorization method, we need first to convert the Laurent polynomial $P(z)$ to the equivalent polynomial $z^N P(z)$ in regular form.

$$P(z) = \prod_{i=0}^{N}(z - z_i)\prod_{j=0}^{N}(z^{-1} - z_j)$$

$$z^N P(z) = \frac{1}{2}a_N \prod_{i=0}^{2N}(z - z_i)$$

with $a_N \neq 0$.

This polynomial can be factored using any numerical package that implements the eigenvalue method discussed above.

The factorization will give us the roots of P. Our knowledge about P will help us to discriminate between the different types of roots:

i. Since P has real coefficients, the complex roots will occur in pairs that are conjugate (i.e., if $z_k = x + iy$ is a root, then $\overline{z_k} = x - iy$ is also a root). This ensures that, in the orthogonal case, we will be able to compute the "square root" of m_0.
ii. The coefficients of P are also symmetric. With real symmetric coefficients a_k, we have $P(z) = P(1/z)$. If z_i is a root, then $1/z_i$ is also a root. Thus, when z_i is inside the unit circle, $1/z_i$ is outside (roots with $|z_i| = 1$, on the unit circle have even multiplicity).

In summary, P has $2N$ roots. The complex roots come in quadruplets $z_j, z_j^{-1}, \overline{z_j}, \overline{z_j^{-1}}$, and the real roots come in duplets r_l, r_l^{-1}. Therefore, regrouping these types of roots, we have

$$z^N P(z) = \frac{1}{2}a_N \prod_{j=1}^{J}(z - z_j)\left(z - \frac{1}{z_j}\right)(z - \overline{z_j})\left(z - \frac{1}{\overline{z_j}}\right)$$
$$\prod_{k=1}^{K}(z - z_k)^2(z - \overline{z_k})^2 \prod_{l=1}^{L}(z - r_l)\left(z - \frac{1}{r_l}\right).$$

Once we obtain the $2N$ roots of $z^N P(z)$, it is necessary to separate then, assigning factors M to the polynomial $m_0(z)$, and the remaining $2N - M$ factors to the polynomial $\overline{m_0(z)}$, (or $\tilde{m}_0(z)$). In the orthogonal case, one factor is the transpose of the other, but in the biorthogonal case, this restriction is not imposed, and the dual scaling functions can be different. Note that, the polynomial $m_0(z)$ with degree N is not unique even for orthogonal scaling functions. In the next two chapters we will discuss the rules for separating the roots of P, and effects of different choices.

The function m_0 is completely determined after we assign a set of roots from P to it. But, we still don't have the coefficients (h_n) of m_0, since it is in product form. Given the roots z_k of the degree N polynomial

$$m_0(z) = c\prod_{k=1}^{N}(z - z_k) \tag{10.6}$$

the coefficients h_n can be computed with an iterated convolution. Then, we normalize the coefficients multiplying them by a scaling constant.

The coefficients (h_n) give the filter m_0, and from them we get the coefficients (g_n) of the filter m_1, using (9.27).

Now, let's summarize the steps required to generate a scaling and wavelet functions using a filter design methodology:

1. Choose a Polynomial P satisfying Condition 1. Generate its coefficients a_k using an explicit formula (see Chaps. 11 and 12).
2. Find the roots z_k of P by a factorization method.
3. Separate the roots z_k into two sets. Assign one set of roots to the polynomial factor m_0 and the other set of roots to $\overline{m_0}$ (or to $\widetilde{m_0}$).
4. Compute the coefficients h_n of m_0 from its roots, using iterated convolution.
5. Obtain the coefficients g_n of m_1 from h_n.

10.1.6 Example (Haar Wavelet)

A concrete example will illustrate the wavelet design methodology.

We choose the following degree 1 polynomial:

$$P(z) = \frac{1}{4}z^{-1} + \frac{1}{2} + \frac{1}{4}z\,.$$

This is the lowest degree polynomial which satisfies Condition 1

$$P(z) + P(-z) = 1\,.$$

We convert P to regular form, multiplying by z

$$zP(z) = \frac{1}{4}\left(1 + 2z + z^2\right)\,.$$

This polynomial has one root at $z = -1$ with multiplicity 2. Therefore P is factored into

$$\begin{aligned} P(z) &= \frac{1}{2}z^{-1}\,(z+1)\quad \frac{1}{2}(z+1) \\ &= \left(\frac{1+z^{-1}}{2}\right)\left(\frac{1+z}{2}\right). \end{aligned}$$

The low-pass filter is

$$m_0(\omega) = \frac{1}{2} + \frac{1}{2}e^{-i\omega} = \frac{1}{\sqrt{2}}\left(\frac{1}{\sqrt{2}} + \frac{1}{\sqrt{2}}e^{-i\omega}\right)$$

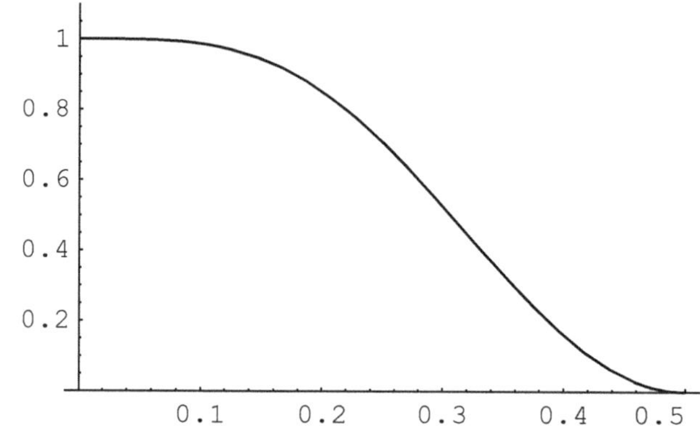

Fig. 10.2 Haar filter m_0

with coefficients $h_0 = h_1 = 1/\sqrt{2}$ which gives the Haar scaling function

$$\phi(x) = \sqrt{2}\left(\frac{1}{\sqrt{2}}\phi(2x) + \frac{1}{\sqrt{2}}\phi(2x-1)\right) = \phi(2x) + \phi(2x-1)\ .$$

Figure 10.2 shows a plot of $m_0(\omega)$.

Notice that, since

$$m_0(\omega) = \frac{1+e^{-i\omega}}{2} = \left(\frac{e^{-i\omega/2}+e^{i\omega/2}}{2}\right)e^{-i\omega/2} = \cos(\omega/2)e^{-i\omega/2}$$

the scaling function in frequency domain $\hat{\phi}$ is the sinc function

$$\hat{\phi}(\omega) = \frac{1}{\sqrt{2\pi}}\prod_{j=1}^{\infty}\cos(2^{-j}\omega/2)e^{-i2^{-j}\omega/2} = \frac{1}{\sqrt{2\pi}}e^{-i\omega/2}\frac{\sin(\omega/2)}{\omega/2}$$

which is the Fourier transform of the box function $\phi(x) = 1$ for x in the interval $[0, 1]$ (this is the Haar scaling function, as expected). A derivation of this formula can be found in [20], p. 211. Figure 10.3 shows a plot of $|\hat{\phi}(\omega)|$.

The coefficients of the high-pass filter m_1 are given by $g_n = (-1)^n h_{1-n}$

$$g_0 = \frac{1}{\sqrt{2}},\quad g_1 = -\frac{1}{\sqrt{2}}\ .$$

The function

$$\psi(x) = \sqrt{2}\left(\frac{1}{\sqrt{2}}\phi(2x) - \frac{1}{\sqrt{2}}\phi(2x-1)\right) = \phi(2x) - \phi(2x-1)$$

is the Haar wavelet.

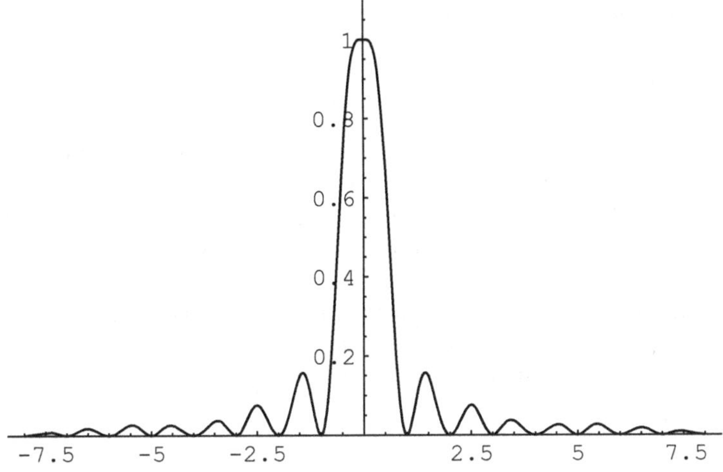

Fig. 10.3 Magnitude of the frequency response of $\hat{\phi}(\omega)$ for the Haar function

10.2 Properties of Wavelets

In the last section we have seen how to synthesize wavelets in the frequency domain. The procedure was based on the factorization of a polynomial $P(\omega)$. In this section we will investigate how the properties of the scaling and wavelet functions relate to characteristics of P. This connection will reveal the "design variables" that we can control to generate wavelets with desired properties.

10.2.1 Orthogonality

The multiresolution analysis, by definition, leads to orthogonal scaling functions and wavelets. In this way, the functions $\{\phi_{j,n}; n \in \mathbb{Z}\}$, and $\{\psi_{j,n}; n \in \mathbb{Z}\}$ are, respectively, orthonormal basis of the subspaces V_j and W_j, for all $j \in \mathbb{Z}$. The subspaces W_j are mutually orthogonal, and the projection operators produce optimal approximations in the $\mathbf{L}^2(\mathbb{R})$ sense.

Orthogonality is very a desirable property, but it imposes severe restrictions on the candidate scaling and wavelet functions. We will see in Chap. 12 that orthogonality can be replaced by biorthogonality, which is less restrictive, giving us more freedom to design wavelets.

10.2.2 Support of ϕ and ψ

In this chapter we have been concerned only with compactly supported basis functions.

The support width of the scaling and wavelet functions is determined, respectively, by the length of the coefficient sequences (h_n) and (g_n). More precisely, if $h_n = 0$ for $n < 0,\ n > N$, then the support of ϕ is the interval $[0, N]$. When ψ is defined as in (9.28), then the support of ψ is the interval $[-M, M+1]$, where $M = (N-1)/2$.

The length of the coefficient sequences (h_n) and (g_n) is given by the degree of the polynomial $m_0(z)$. Consequently, the degree of the product polynomial $P(z)$, determines the support width of the associated basis functions.

10.2.3 Vanishing Moments and Polynomial Reproduction

Two related properties of a multiresolution analysis are: the vanishing moments of the wavelet functions; and the ability of the scaling functions to reproduce polynomials. These are probably the most important properties in the design of wavelets.

Given a mother wavelet function $\psi(x)$ with p vanishing moments, that is

$$\int x^{\ell}\psi(x)dx = 0 \tag{10.7}$$

for $\ell = 0, 1, \ldots, p-1$. Then, the following equivalent properties are verified:

- The wavelets $\psi(x-k)$ are orthogonal to the polynomials $1, x, \ldots, x^{p-1}$.
- The combination of scaling functions $\phi(x-k)$ can reproduce exactly the polynomials $1, x, \ldots, x^{p-1}$.

These properties are important in the rate of convergence of wavelet approximations of smooth functions and in singularity detection using wavelets.

Vanishing moments are also a necessary (but not sufficient) condition for a wavelet to be $p-1$ times continuously differentiable, i.e. $\psi \in C^{p-1}$.

In order to make the connection between the number of vanishing moments of a wavelet and the condition imposed on the filter function $m_1(\omega)$, we note that Eq. (10.7) implies that

$$\left.\frac{d^{\ell}}{d\omega^{\ell}}\hat{\psi}\right|_{\omega=0} = 0 \tag{10.8}$$

for $\ell < p$. (This can be verified expanding ψ in a Taylor series.) But, we know from (9.26) that $\hat{\psi}(\omega) = e^{-i\omega/2}\overline{m_0(\omega/2+\pi)}\hat{\phi}(\omega/2)$, and also from (9.5) that $\hat{\phi}(0) \neq 0$, this means that m_0 is $p-1$ times differentiable in $\omega = \pi$. Furthermore

$$\frac{d^\ell}{d\omega^\ell} m_0 \bigg|_{\omega=\pi} = 0 \tag{10.9}$$

for $\ell < p$.

As a consequence of the above, m_0 must have a zero of order p at $\omega = \pi$. Therefore, m_0 is of the form

$$m_0(\omega) = \left(\frac{1 + e^{i\omega}}{2}\right)^p Q(\omega) \tag{10.10}$$

with $m_0 \in C^{p-1}$ and $Q \in C^{p-1}$.

We already knew that a wavelet should integrate to zero, i.e. it should have at least one vanishing moment. We also knew that m_0 should have at least one zero at π, i.e., it should be of the form (10.10) with $p >= 1$.

The zero at π from m_0 produces factor $(\frac{1+z}{2})^{2p}$ in the polynomial P.

10.2.4 Regularity

Smooth basis functions are desired in applications where derivatives are involved. Smoothness also corresponds to better frequency localization of the filters.

The local regularity of a function at a point x_0 can be studied using the notion of Lipschitz continuity. A function $f(x)$ is C^α at x_0, with $\alpha = n + \beta$, $n \in \mathbb{N}$ and $0 < \beta < 1$, if f is n times continuously differentiable at x_0 and its nth derivative is Lipschitz continuous with exponent β.

$$|f(x)^{(n)} - f(x_0)^{(n)}| \leq C|x - x_0|^\beta \; . \tag{10.11}$$

For global regularity, the above must hold for all x_0.

Typically, compactly supported scaling functions and wavelets are more regular in some points than in others. They also have non-integer Lipschitz exponents.

The regularity of scaling functions and wavelets is difficult to determine. The techniques for this investigation involve either the estimation of the decay of $\hat{\phi}$ in the frequency domain, or the estimation of the convergence rate of the recursive construction of ϕ in the spatial domain. The first class of techniques is more suitable for a global analysis, while the second class of techniques is better for a local analysis.

The regularity of ϕ and ψ is related with the number of vanishing moments of ψ, and consequently is related with the multiplicity p of the zeros at π of m_0. The regularity is at most $p - 1$ if we consider integer exponents, and at most $p - 1/2$ if we consider fractional exponents. We remark that in many cases, the regularity is much smaller than $p - 1$ because of the influence of the factor Q.

Another important observation is that, for a fixed support width of ϕ, the choice of factorizations of P that leads to maximum regularity is different from the choice with maximum number of vanishing moments for ψ.

10.2.5 Symmetry or Linear Phase

Symmetry is important in many applications of wavelets, such as image processing and computer vision. This property can be exploited in the quantization of images for compression, and in the boundary processing of finite signals using symmetric extension (i.e. periodization by reflection of the data).

The symmetry of the basis functions implies in symmetric filters. It is easy to see from (9.1) that ϕ will be symmetric only if the coefficients (h_n) are symmetric.

Symmetric filters are called *linear phase* filters. This is because filters with symmetric coefficients have a linear phase response.

The phase angle (or argument) of a complex number $z = a + bi$ is the angle between the vector (a, b) and the horizontal axis. The tangent of this angle is the ratio of the imaginary and real parts of z:

$$\tan(\omega) = \frac{\Im(z)}{\Re(z)} = \frac{b}{a} \, . \tag{10.12}$$

The same definition extends to a complex function $m(z)$. This notion becomes more evident if we use the polar representation of a complex number $z = |z|e^{i\theta(\omega)}$, where $|z|$ is the magnitude of z and $\theta(\omega)$ is the phase angle of z.

A filter function $m(\omega) = \sum_k a_k e^{ik\omega}$ is said to have linear phase if $\theta(\omega)$ is of the form $K\omega$ where K is a constant (i.e., it is linear in ω). This means that $m(\omega)$ can be written as

$$m(\omega) = ce^{iK\omega} M(\omega) \tag{10.13}$$

where c is a complex constant, and M is a real valued function of ω, not necessarily 2π-periodic. Linear phase filters have symmetric or antisymmetric coefficients around the central coefficient $N/2$. Therefore, $h_k = h_{N-k}$ for symmetric filters and $h_k = -h_{N-k}$ for antisymmetric filters. Note that, when filter coefficients a_k are symmetric, or anti-symmetric, the phase is linear because terms with same coefficients can be grouped. For example, if the filter is symmetric and has an even number of coefficients

$$m(\omega) = \sum_{k=0}^{N/2} h_k(e^{-ik\omega} + e^{-i(N-k)\omega}) = e^{-i\frac{N}{2}\omega} \left(\sum_{k=0}^{N/2} h_k(e^{ik\omega} + e^{-ik\omega}) \right)$$

$$= e^{-i\frac{N}{2}\omega} \left(\sum_{k=0}^{N/2} 2h_k \cos(k\omega) \right) \, .$$

We can see that in this case the phase is $N/2\omega$. Similar expressions hold for the case of symmetric filters with odd number of coefficients, as well as for antisymmetric filters.

We can conclude from the above discussion that, in order to obtain symmetric (antisymmetric) scaling functions and wavelets, it is necessary to factor the polynomial P into symmetric polynomials m_0 (and $\widetilde{m_0}$). Therefore, in the factorization of P the roots z_j and z_j^{-1} must stay together. We remark that, although P is a symmetric polynomial, it obviously can be factorized into two polynomials that are not symmetric.

In the next chapter, we will demonstrate that symmetry and orthogonality are incompatible properties, except for the case of Haar wavelets. This is because orthogonal filters require that the roots z_j and z_j^{-1} are separated and assigned one to m_0 and the other to $\overline{m_0}$.

Nonetheless, despite the orthogonality restriction, it is possible to design filters that are close to linear phase and result in basis functions that are least asymmetric. This can be accomplished by selecting from each pair of roots z, z^{-1}, the one which contributes the least to nonlinearity of the filter phase. For this purpose, we have to compute the nonlinear contribution of the factors $(z - z_k)$ and $(z - z_k^{-1})$ to the total phase of m_0.

10.2.6 Other Properties

There are several other properties that are relevant to applications, but are only achievable by some types of wavelets. Below we discuss a few of them.

- **Analytic Form**: The analytic expression for the scaling function and wavelet is, in general, not available. These functions are defined indirectly through the filter coefficients (h) and (g). Nonetheless, the definition of the scaling function and wavelet in analytic form is very useful in many applications. One important example of wavelet basis with closed form analytic expression is the family of B-Spline wavelets.
- **Interpolation:** The interpolation property makes it trivial to find the values of the scaling coefficients from samples of the function. When the scaling function $\phi(k) = \delta_k$, for $k \in \mathbb{Z}$, the coefficients $c_k^j = \langle f, \phi_j \rangle$ of the projection $\mathrm{Proj}_{V_j}(f)$ of f on V_j, are just the values $f(2^j k)$ of f sampled on a grid with spacing 2^{-j}.
- **Rational Coefficients:** Filter coefficients that are rational numbers make computer implementation more efficient and precise. It is even better if the coefficients are dyadic rational numbers. In this case, division by a power of two corresponds to shifting bits.

10.3 Classes of Wavelets

From our discussion of the various properties of wavelets, we can conclude that it is not possible to achieve all of them at the same time. The reason is that different properties may be incompatible with each other, and imply in conflicting requirements on the filter function m_0. For example, on one hand, the support width is proportional to the number of filter coefficients and consequently small support requires a low-degree m_0. On the other hand, the number of vanishing moments depends on the multiplicity of the zero at π of m_0, and consequently a large number of vanishing moments requires a high degree m_0.

To make things even more complicated, there is a great interdependency among the different requirements. For instance, m_0 is of the form $1/2(1 - e^{i\omega})^n Q(\omega)$. While the first factor controls the vanishing moments, the second factor is necessary to guarantee orthogonality. If we increase the degree of the first, we also have to increase the degree of the second.

Naturally, as in any design process, there is always a trade-off between different properties relative to their importance in a given application.

In general terms, there are two main classes of wavelets that combine some of the above properties in the best way.

10.3.1 Orthogonal Wavelets

Orthogonal wavelets are the best choice in some cases. Unfortunately, except for the Haar case, orthogonal wavelets cannot be symmetric.

It is possible to design orthogonal wavelets with more or less asymmetry.

Other properties that are important for orthogonal wavelets are: support width, number of vanishing moments, and regularity.

In Chap. 11, we will discuss how to generate compactly supported wavelets, and how to control the various properties.

10.3.2 Biorthogonal Wavelets

Biorthogonal wavelets have most of the qualities of orthogonal wavelets, with the advantage of being more flexible.

There are many more biorthogonal wavelets than orthogonal ones. For these reasons, they make possible a variety of design options and constitute class of wavelets most used in practical applications.

Biorthogonal wavelets can have symmetry. They are associated with perfect analysis/reconstruction filter banks.

In Chap. 12 we will discuss how to generate biorthogonal wavelets.

10.4 Comments and References

The framework for wavelet design in the frequency domain has been developed since the early days of wavelet research by Yves Meyer, [43], and Stephane Mallat, [36], among others. The initial methods investigated could only produce orthogonal wavelets with infinite support. The major breakthrough was made by Ingrid Daubechies in 1988, [18].

Gilbert Strang presents in his book, [55], a complete and detailed framework for generating wavelets using the filter design methodology.

Carl Taswell developed a computational procedure to generate wavelet filters using the factorization of the polynomial P, [60].

Chapter 11
Orthogonal Wavelets

In this chapter we will investigate the construction and design of compactly supported orthogonal wavelets. We will derive a closed form expression for the polynomial P, introduced in the previous chapter, and we will show how to factor P in order to generate orthogonal wavelets with different properties.

11.1 The Polynomial P

We saw in the previous chapter that the trigonometric polynomial $P(z)$, with $z = e^{i\omega}$ is the key element in the construction of compactly supported wavelets. There, we described a method to synthesize wavelets by factoring P as a product of two polynomials. But, in order to apply this method, we need first an explicit expression for P, i.e. a formula to compute the coefficients a_k of $P(e^{i\omega}) \sum_{k=-n}^{n} a_k e^{ik\omega}$.

To find a closed form expression for P we recall the two conditions that P must satisfy in the context of orthogonal filter banks.

- P is a reciprocal product polynomial

$$P(e^{i\omega}) = |m_0(e^{i\omega})|^2 \tag{11.1}$$

  where m_0 has a zero of order n at $\omega = \pi$ (or $e^{i\omega} = -1$).
- P is a halfband filter

$$P(e^{i\omega}) + P(e^{i\omega+\pi}) = 1 \ . \tag{11.2}$$

We need to find a polynomial P of the form (11.1), such that Eq. (11.2) is true. We will address these two conditions one at time, and we will combine them afterwards to obtain an expression for P.

J. Gomes, L. Velho, *From Fourier Analysis to Wavelets*,
IMPA Monographs 3, DOI 10.1007/978-3-319-22075-8_11

11.1.1 P as a Product Polynomial

The polynomial $P(z) = |m_0(z)|^2$ is generated from m_0, which in turn is of the form

$$m_0(e^{i\omega}) = \left(\frac{1+e^{i\omega}}{2}\right)^n Q_n(e^{i\omega}) . \tag{11.3}$$

The factor $(1 + e^{i\omega})^n$ is a consequence of the fact that m_0 has a zero of order n at π.

We get an expression for P by substituting Eq. (11.3) into (11.1)

$$P(e^{i\omega}) = |m_0(e^{i\omega})|^2 \tag{11.4}$$

$$= m_0(e^{i\omega})\,\overline{m_0(e^{i\omega})} \tag{11.5}$$

$$= m_0(e^{i\omega})\, m_0(e^{-i\omega}) \tag{11.6}$$

$$= \left(\frac{1+e^{i\omega}}{2}\right)^n \left(\frac{1+e^{-i\omega}}{2}\right)^n |Q_n(e^{i\omega})|^2 . \tag{11.7}$$

Since $|1 + e^{-i\omega}|^2/2 = (1 + \cos\omega)/2$, the condition of n zeros at π appears in P as

$$P(\omega) = \left(\frac{1+\cos(\omega)}{2}\right)^n |Q_n(e^{i\omega})|^2 . \tag{11.8}$$

The factor $|Q_n(e^{i\omega})|^2$ in P gives extra degrees of freedom to satisfy the halfband property expressed in Eq. (11.2).

A question comes up at this point. Is it possible to factor every polynomial $P(e^{i\omega}) \geq 0$ into a product $|m_0(e^{i\omega})|^2$? The answer is yes. This is guaranteed by the Riesz Lemma.

Lemma 11.1 (Riesz). *If $A(\omega) \geq 0$ is a trigonometric polynomial of degree M, invariant under the substitution $\omega \to -\omega$, A has to be of the form*

$$A(\omega) = \sum_{m=0}^{M} a_m \cos m\omega$$

with $a_m \in \mathbb{R}$. Then there exists a polynomial B of order M

$$B(\omega) = \sum_{m=0}^{M} b_m e^{im\omega}$$

with $b_m \in \mathbb{R}$, such that $|B(\omega)|^2 = A(\omega)$.

A proof of this lemma can be found in [20], p. 172.

The above result guarantees that a factorization of P using the method described in the previous chapter is possible. In fact, the proof is constructive and follows the steps necessary for the spectral factorization. Note that P is a polynomial in $\cos\omega$ as enunciated by the lemma.

11.1.2 P and the Halfband Property

We derived above a partial expression for P. We found that P has to be decomposed into two factors $P(\omega) = p_1(\omega)q_1(\omega)$, where $p_1(\omega) = (1+\cos\omega)^n/2$ provides the zeros at π and q_1 are chosen such that the halfband property is verified.

To relate P and the halfband filter property in Eq. (11.2), we use the following theorem:

Theorem 11.1 (Bezout). *If p_1 and p_2 are two polynomials with no common zeros, then there exist unique polynomials q_1 and q_2 such that*

$$p_1(y)q_1(y) + p2(y)q_2(y) = 1 \tag{11.9}$$

where $p1$, p_2, q_1, q_2, are of degree n_1, n_2, n_2-1, n_1-1, respectively.

A proof of this theorem can be found in [20], p. 169.

For reasons that will become clear later, let's apply Bezout's theorem to the particular case where $p_1(y) = (1-y)^N$ and $p_2(y) = y^N$. The theorem says that there exist unique polynomials q_1, q_2 of degree $\le N-1$, such that

$$(1-y)^N q_1(y) + y^N q_2(y) = 1 \ . \tag{11.10}$$

When we substitute y for $1-y$, Eq. (11.10) becomes

$$(1-y)^N q_2(1-y) + y^N q_1(1-y) = 1 \ . \tag{11.11}$$

Because q_1 and q_2 are unique, necessarily $q_1(y) = q_2(1-y)$, and also $q_2(y) = q_1(1-y)$.

This allows us to find an explicit formula for q_1.

$$\begin{aligned} q_1(y) &= (1-y)^{-N}[1 - y^N q_1(1-y)] \\ &= (1-y)^{-N} - y^N R(y) \\ &= \sum_{k=0}^{N-1} \binom{N+k-1}{k} y^k - y^N R(y) \end{aligned} \tag{11.12}$$

where we expanded the first N terms of the Taylor series for $(1-y)^{-N}$. Since the degree of $(q_1) \le N-1$, q_1 is equal to its Taylor expansion truncated after N terms.

Therefore, we can drop $y^N R(y)$ and make $q_1(y) = (1-y)^{-N}$. This gives a closed form expression for q_1, which is the unique lowest degree solution for (11.10).

$$(1-y)^N B_n(y) + y^N B_n(1-y) = 1 \tag{11.13}$$

where $B_n(y)$ is the binomial series for $(1-y)^{-N}$, truncated after N terms

$$B_n(y) = \sum_{k=0}^{N-1} \binom{N+k-1}{k} y^k . \tag{11.14}$$

Higher degree solutions for (11.10) also exist with $q_1(y) = B_n(y) - y^N R(y)$, provided that $R(y)$ satisfies $R(y) - R(1-y) = 0$.

11.1.3 The Expression of P

Now we put together the two results above to obtain an expression for P. So far we know the following:

i. P is of the form $P(\omega) = \left(\frac{1+\cos(\omega)}{2}\right)^n |Q_n(e^{i\omega})|^2$
ii. A polynomial of the form $P(y) = (1-y)^n B_n(y)$ has the halfband property.

To relate (i) and (ii) we make a change of variables

$$\frac{e^{i\omega} + e^{-i\omega}}{2} = \cos(\omega) = 1 - 2y$$

which gives

$$y = \frac{1-\cos(\omega)}{2} \quad \text{and} \quad 1 - y = \frac{1+\cos(\omega)}{2} .$$

Therefore, we can write P as

$$P(\omega) = \left(\frac{1+\cos(\omega)}{2}\right)^n \sum_{k=0}^{n-1} \binom{n+k-1}{k} \left(\frac{1-\cos(\omega)}{2}\right)^k . \tag{11.15}$$

It is clear that B_n is the solution for $|Q_n|^2$ which makes P satisfy the halfband property.

Making another change of variables we obtain an expression for P in terms of the complex variable $z = e^{i\omega}$.

$$\frac{1+\cos(\omega)}{2} = \left(\frac{1+e^{i\omega}}{2}\right)\left(\frac{1+e^{-i\omega}}{2}\right) = \left(\frac{1+z}{2}\right)\left(\frac{1+z^{-1}}{2}\right)$$
$$\frac{1-\cos(\omega)}{2} = \left(\frac{1-e^{i\omega}}{2}\right)\left(\frac{1-e^{-i\omega}}{2}\right) = \left(\frac{1-z}{2}\right)\left(\frac{1-z^{-1}}{2}\right)$$

Substituting this into Eq. (11.15) we arrive at the closed form expression for P in the z-domain

$$P(z) = \left(\frac{1+z}{2}\right)^n \left(\frac{1+z^{-1}}{2}\right)^n \sum_{k=0}^{n-1} \binom{n+k-1}{k} \left(\frac{1-z}{2}\right)^k \left(\frac{1-z^{-1}}{2}\right)^k . \tag{11.16}$$

Note that to produce P in the form of a Laurent polynomial $P(z) = \sum_{k=-n}^{n} a_k z^k$, it is necessary to expand the factors in each term, and then collect the resulting terms with same exponent. This would be required to factorize $P(z)$ directly, but we will see that it is more efficient to compute the roots of $P(y) = (1-y)^n B_n(y)$ and make the change of variables afterwards.

11.1.4 The Factorization of P

The two forms of the polynomial $P(z)$ and $P(y)$ are related by the change of variables

$$(z + z^{-1})/2 = 1 - 2y . \tag{11.17}$$

We write this as a composition of conformal maps $(f \circ g)(z) = y$, where f is the Joukowski transformation [42],

$$x = f(z) = \frac{z + z^{-1}}{2}$$
$$z = f^{-1}(x) = x \pm \sqrt{x^2 - 1}$$

and g is an affine transformation

$$y = g(x) = (1-x)/2$$
$$x = g^{-1}(y) = 1 - 2y .$$

The expressions for the change of variables are

$$y = g(f(z)) = (1 - (z + z^{-1})/2)/2 \tag{11.18}$$
$$z = f^{-1}(g^{-1}(y)) = 1 - 2y \pm \sqrt{(1-2y)^2 - 1} . \tag{11.19}$$

Notice that this change of variables in (11.19) associates two values of z for each value of y. Equation (11.18) is, in fact, a quadratic equation for z. It is clear from (11.17) that one z is inside the unit circle and the other $1/z$ is outside.

The polynomial $P(y) = (1-y)^n B_n(y)$, from Subsect. 11.1.3, has degree $2n-1$, but because of the change of variables, it will result in a polynomial $P(z)$ of degree $4n-2$. Thus, $P(z)$ will have $4n-2$ roots. From these, $2n$ roots come from the first factor $(1-y)^n$, and correspond to the multiple root at $z = -1$. The remaining $2n-2$ complex roots come from the binomial factor $B_n(y)$.

The polynomial $|Q_n(e^{i\omega})|^2$ is a reciprocal polynomial with real coefficients. Thus, its roots are in reciprocal and complex conjugate pairs. The change of variables $z = f^{-1}(g^{-1}(y))$ yields a doubly valued solution with reciprocal pairs $\{z, z^{-1}\}$. The polynomial $|Q_n(z)|^2$, expressed in regular form, can be factored by regrouping these pairs as complex quadruplets $\{z, z^{-1}, \bar{z}, \bar{z}^{-1}\}$ and real duplets $\{r, r^{-1}\}$

$$|Q_n(z)|^2 = \prod_{i=1}^{K} U(z; z_i) \prod_{j=1}^{L} V(z; r_j) \tag{11.20}$$

$$U(z; z_i) = (z - z_i)(z - z_i^{-1})(z - \bar{z}_i)(z - \bar{z}_i^{-1}) \tag{11.21}$$

$$U(z; r_j) = (z - r_j)(z - r_j^{-1}) \tag{11.22}$$

where $K = (n-1)/2$ and $L = (n-1) \mod 2$.

The factored polynomial $z^{2n-1}P(z)$, in regular form, is obtained by including the multiple zeros at $z = -1$.

$$z^{2n-1}P(z) = (z+1)^{2n} \prod_{i=1}^{K} U(z; z_i) \prod_{j=1}^{L} V(z; r_j) \tag{11.23}$$

The filter function $m_0(z)$ can be computed from this factorization by noting that

$$P(z) = |m_0(e^{i\omega})|^2 = m_0(e^{i\omega})\overline{m_0(e^{i\omega})} = m_0(e^{i\omega})m_0(e^{-i\omega}) = m_0(z)m_0(z^{-1}) \, .$$

$$m_0(z) = c \prod_{i=1}^{K} (z - z_i)(z - \bar{z}_i) \prod_{j=1}^{L} (z - r_j) \tag{11.24}$$

where c is a constant, and z_i and r_j are selected from the elements of each reciprocal pair $\{z, z^{-1}\}$.

In consequence of the special structure of P, it is more efficient to compute the $2n-1$ roots of $P(y)$ first, and using a change of variables, generate the $4n-2$ roots of $P(z)$. The roots of m_0 are then selected from those roots. Therefore, the spectral factorization of m_0 is done in two steps:

1. Find the $n-1$ roots of $B_n(y)$, and make a change of variables to produce the $2n-2$ roots of $|Q_n(z)|^2$. From these, select appropriately $n-1$ roots and assign them to m_0.
2. Include the n roots at $z=-1$ from the factor $\left(\frac{1+z}{2}\right)^n$ of $\left(\frac{1+z}{2}\right)^n\left(\frac{1+z^{-1}}{2}\right)^n$, and assign them also to m_0.

The orthogonal filter function $m_0(z)$ will have these $2n-1$ roots.

11.1.5 Analysis of P

The family of orthogonal filters generated from the polynomial $P(y)=(1-y)^n B_n(y)$ is called *maxflat* in the filter design literature. This is because the factor $(1-y)^n$ gives a maximum number of zeros at $y=1$ (or $z=-1$). Consequently, the filter response of the filter is maximally flat at $\omega=\pi$.

In fact, the polynomial P of degree $2n-1$ is determined from n conditions at $y=1$ and at $y=0$. $P(y)$ is the unique polynomial with $2n$ coefficients which satisfies these n conditions at the endpoints, i.e. $P(y)$ and its first $n-1$ derivatives are zero at $y=0$ and at $y=1$, except that $P(0)=1$.

Figure 11.1 shows a graph of $P(y)$ for $n=2$ and 4. Observe that P decreases monotonically from $P(0)=1$ to $P(1)=0$. The slope of P at $y=\frac{1}{2}$ is proportional to $\sqrt{n}$. As n increases, P becomes simultaneously steeper at the midpoint and flatter at both endpoints.

This property is very important because it means that P (and its "square root" m_0) as a filter will preserve low frequencies (near $\omega=0$) and will cut off high frequencies (near $\omega=\pi$).

Gilbert Strang studied the asymptotic behavior of the zeros of $P(y)$ as $n\to\infty$. In [54], there is an insightful analysis of the factors $B_n(y)$ in $P(y)=(1-y)^n B_n(y)$, and $|Q_n(z)|^2$ in $P(z)=|(\frac{1+z}{2})^n|^2|Q_n(z)|^2$. We recall that $P(y)$ and $P(z)$ are related by

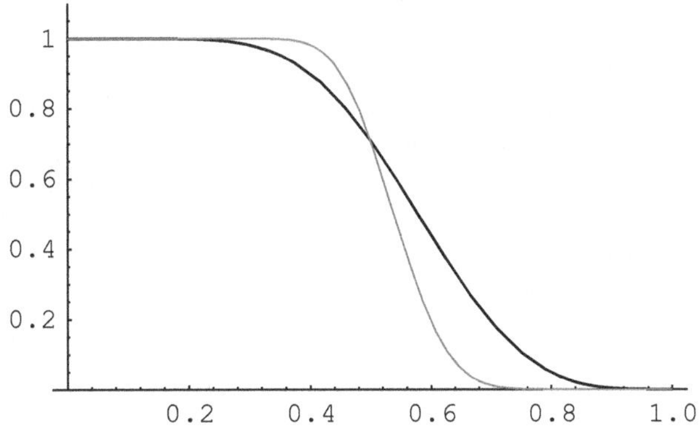

Fig. 11.1 P - the maxflat polynomial generator

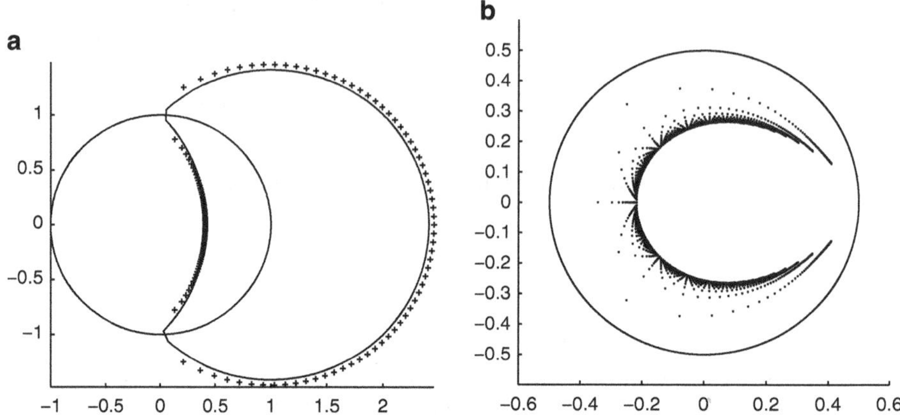

Fig. 11.2 Limit curves for the roots of $B_n(y)$ and $|Q_n(z)|^2$ (from [54])

the change of variables $(1-2y) = (z+z^{-1})/2$. Also, it is shown that, as $n \to \infty$, the roots of $B_n(y)$ approach the limiting curve $|4y(1-y)| = 1$, whereas the roots of $|Q_n(z)|^2$ approach the curve $|z - z^{-1}| = 2$.

Figure 11.2 shows a plot of the graphs of these two curves in the complex plane. Note that, because the change of variables yields a doubly valued solution, the curve in y actually corresponds to two curve segments $|z+1| = \sqrt{2}$ and $|z-1| = \sqrt{2}$. These curves meet at $z = \pm i$, which correspond to $y = 1/2$.

11.2 Examples of Orthogonal Wavelets

In this section we show how to generate, from the polynomial P, particular families of orthogonal wavelets.

We have seen in the previous section that the factorization of $P(z)$ produces roots that can be grouped as quadruplets of complex zeros $(z-z_i)(z-z_i^{-1})(z-\bar{z}_i)(z-\bar{z}_i^{-1})$ and duplets of real zeros $(z-r_j)(z-r_j^{-1})$. In the spectral factorization of $P(z) = |m_0(z)|^2$, we have to choose from each pair of reciprocal roots $\{z_k, z_k^{-1}\}$ of P, either z_k or z_k^{-1} as a root of m_0. Also, if we require that m_0 is a polynomial with real coefficients, complex conjugate pairs must stay together. Therefore, in this case, we can select either $\bar{z}_k$ with z_k or $\bar{z}_k^{-1}$ with z_k^{-1}.

Even with these restrictions for separating the roots of $P(z)$, we have many choices to Taylor m_0 according to design constraints. Typically, for P of degree m, there are $2^{\frac{m}{4}}$ different choices for m_0. Let's now take a look at the most important ones.

11.2.1 Daubechies Extremal Phase Wavelets

The simplest systematic choice for designing orthogonal wavelets produces the so-called extremal phase wavelets. This corresponds to selecting always the roots inside (or outside) the unit circle, with $|z| < 1$ (or $|z| > 1$), from each reciprocal pair of roots $\{z, \frac{1}{z}\}$ of P.

The resulting orthogonal wavelets will have the minimum (or maximum) phase among all compactly supported wavelets of degree m. This corresponds to scaling functions and wavelets that are the most asymmetric basis functions. We remark also that the minimum and maximum phase are complementary choices. They lead to m_0 and the complex conjugate of m_0. Thus, the basis functions are mirror images of each other for these two choices.

Example 16 (Daubechies Wavelets of Order 2). In this case $n = 2$. Therefore $P(y) = (1-y)^2 B_2(y)$, with $B_2(y) = 1 + 2y$.

$$P(y) = (1-y)^2(1+2y) . \tag{11.25}$$

The change of variables $y \to z$ gives

$$P(z) = \left(\frac{1+z}{2}\right)^2 \left(\frac{1+z^{-1}}{2}\right)^2 \frac{1}{2}\left(-z+4-z^{-1}\right) . \tag{11.26}$$

The factor $B_2(y)$ has one root at $y = -1/2$. Equation (11.17) for the change of variables gives the roots of $|Q_2(z)|^2$ at $z = 2 \pm \sqrt{3}$. The factor $(1-y)^2$ will produce the four zeros at $z = -1$ of P. The roots of $P(z)$ include all these zeros, and are $(-1, -1, -1, -1, 2-\sqrt{3}, 2+\sqrt{3})$.

The choice of the root for $Q_2(z)$ inside the unit circle $z = 2 - \sqrt{3}$ leads to the minimum phase m_0 with roots $(-1, -1, 2-\sqrt{3})$. Figure 11.3 shows the roots of P and m_0.

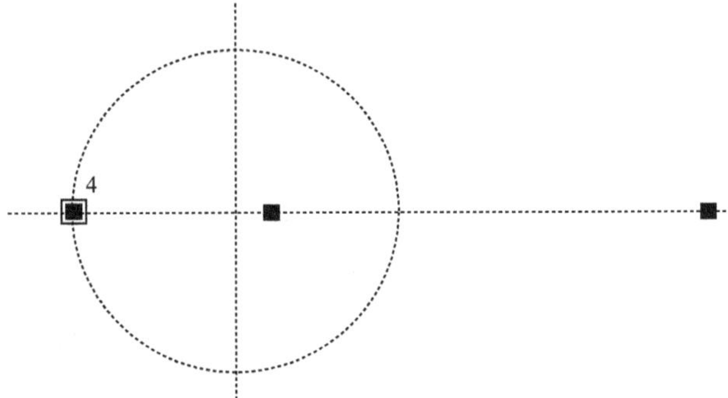

Fig. 11.3 Roots of $P(z)$ with $n = 2$

The filter function m_0 is then

$$m_0(z) = \left(\frac{1+z}{2}\right)^2 Q_2(z) \tag{11.27}$$

$$= c(1+z^{-1})^2(1-(2-\sqrt{3})z^{-1}) \tag{11.28}$$

$$= \frac{1}{4\sqrt{2}}\left((1-\sqrt{3}) + (3+\sqrt{3})z^{-1} + (3-\sqrt{3})z^{-2} + (1-\sqrt{3})z^{-3}\right) \tag{11.29}$$

$$= h_0 + h_1 z^{-1} + h_2 z^{-2} + h_3 z^{-3}\,. \tag{11.30}$$

The coefficients $\{h_n\}$ of m_0 are approximately 0.4830, 0.8365, 0.2241, –0.1294.

Figure 11.4 shows a plot of the Daubechies scaling function $\phi_{D,2}$ and wavelet $\psi_{D,2}$. Figure 11.5 shows the graphs of the Fourier Transform of the Daubechies scaling function $\phi_{D,2}$ and wavelet $\psi_{D,2}$.

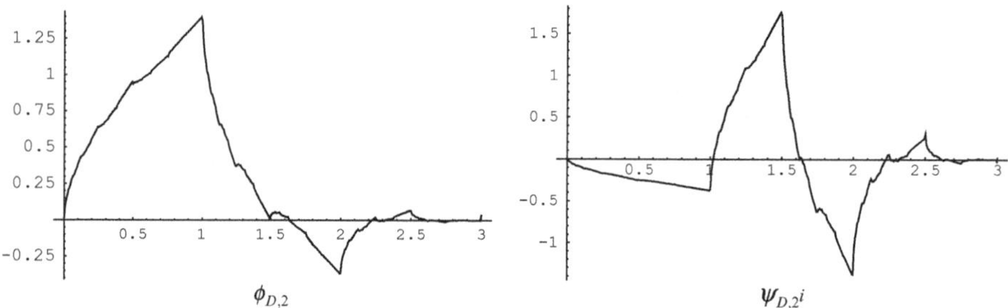

Fig. 11.4 Daubechies (D2) Scaling Function and Wavelet

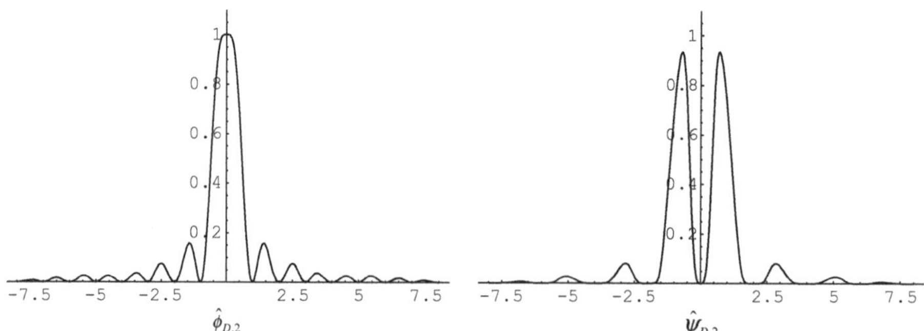

Fig. 11.5 Fourier Transform of (D2) Scaling Function and Wavelet

11.2.2 Minimal Phase Orthogonal Wavelets

The choice of m_0 having all roots with absolute value less (or greater) than one leads to wavelets with a very marked asymmetry.

Symmetry and orthogonality are conflicting requirements for compactly supported wavelets. It can be shown that the Haar basis constitutes the only scaling functions and wavelets with compact support which are orthogonal.

We are not going to demonstrate this result, but it is intuitive to understand, from the requirements on filter m_0, why orthogonality and symmetry are incompatible. Recall from the rules for separating the roots of $P(z)$ that:

- for orthogonal filters, z and z^{-1} must go separately; and
- for symmetric filters z and z^{-1} must stay together.

The above facts imply that a symmetric orthogonal FIR filter $m_0(z)$ can only have two non-zero coefficients. In this case,

$$P(z) = \left(\frac{1+z}{2}\right)\left(\frac{1+z^{-1}}{2}\right),$$

which corresponds to the Haar wavelets with

$$m_0(z) = (1 + z^{-1})/\sqrt{2}\,.$$

Nonetheless, it is possible to construct compactly supported orthogonal wavelets that are less asymmetric than the ones in the previous subsection.

In order to generate the least asymmetric basis functions, we have to select the roots of P such that m_0 is as close as possible to linear phase. Therefore, we need to estimate the contribution of each root to the phase nonlinearity of m_0.

Since m_0 is of the form

$$m_0(\omega) = \left(\frac{1+e^{-i\omega}}{2}\right)^n \prod_\ell (e^{-i\omega} - z_\ell)(e^{-i\omega} - \bar{z}_\ell) \prod_k (e^{-i\omega} - r_k) \tag{11.31}$$

where $z_\ell, \bar{z}_\ell, r_k$ are the roots of m_0 and we have substituted $z = e^{-i\omega}$ in Eq. (11.23). The phase of m_0 can be computed from the phase of each $z_\ell, \bar{z}_\ell$, and r_k.

Since $z_\ell = r_\ell\, e^{-i\alpha_\ell}$ and $\bar{z}_\ell = r_\ell\, e^{i\alpha_\ell}$

$$(e^{-i\omega} - r_\ell\, e^{-i\alpha_\ell})(e^{-i\omega} - r_\ell\, e^{i\alpha_\ell}) = e^{-i\omega}(e^{-i\omega} - 2r_\ell \cos\alpha_\ell + r_\ell^2 e^{i\omega}) \tag{11.32}$$

their phase contribution $\tan\omega = \Im z/\Re z$ is

$$\arctan\left(\frac{(r_\ell^2 - 1)\sin\omega}{(r_\ell^2 + 1)\cos\omega - 2r_\ell\cos\alpha_\ell}\right). \tag{11.33}$$

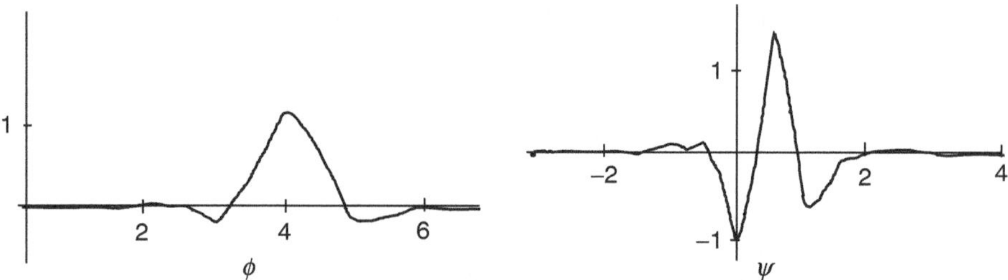

Fig. 11.6 Least Asymmetric Scaling Function and Wavelet, with $n = 4$

Similarly, since

$$(e^{-i\omega} - r_k) = e^{-i\omega/2}(e^{-i\omega/2} - r_k e^{i\omega/2}) \tag{11.34}$$

the phase contribution of r_k is

$$\arctan\left(\frac{(r_k + 1)}{(r_k - 1)} \tan\frac{\omega}{2}\right) . \tag{11.35}$$

To find the polynomial m_0 of degree n with the smallest phase nonlinearity it is necessary to compare all possible combination of roots. In practice, there are $2^{\frac{n}{2}-1}$ choices. For $n = 2$ or 3, there is effectively only one set of basis $\phi_{D,n}$, $\psi_{D,n}$. For $n \geq 4$, we have to evaluate the total phase nonlinearity of m_0 for all $2^{\frac{n}{2}-1}$ solutions.

Figure 11.6 shows the graph of the least asymmetric ϕ and ψ for $n = 4$.

11.2.3 Coiflets

Up to this point we have assumed that m_0 is the unique lowest degree polynomial which satisfy (11.2), i.e. $|m_0(y)|^2 = (1 - y)^n B_n(y)$. We can use a higher degree $|m_0(y)|^2 = (1 - y)^n [B_n(y) - y^n R(y)]$ to provide more freedom in the design o orthogonal wavelets. The price to pay is a wider support of the basis functions for a given number of vanishing moments. The minimum support width $2n - 1$ for ψ with n vanishing moments is achieved when $R \equiv 0$.

Daubechies resorted to such higher degree solutions to construct a family of wavelets named "coifflets," because R. Coiffman requested them motivated by the research in [3]. This basis functions are more regular and symmetric than the ones presented in the previous subsections. More precisely, the scaling functions and wavelets are designed so that:

$$\int \phi(x)dx = 1, \quad \int x^\ell \phi(x)dx = 0 \tag{11.36}$$

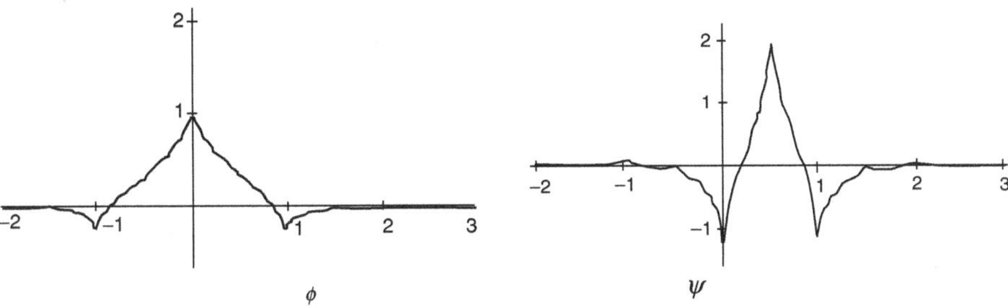

Fig. 11.7 Coifflets of order 4

for $\ell = 1, \ldots, L-1$ and

$$\int x^{\ell}\psi(x)dx = 0 \tag{11.37}$$

for $\ell = 0, \ldots, L-1$.

In other words, both ψ and ϕ have n vanishing moments, except that the scaling function integrates to one. This implies that $\langle f, \phi_{-J,k}\rangle \approx 2^{J/2}f(2^{-J}k)$, making very simple to obtain the fine scale coefficients $\langle f, \phi_{-J,k}\rangle$ from samples of f.

Figure 11.7 shows a plot of the coifflets of order 4.

11.3 Comments and References

The main result for the construction of compactly supported orthogonal wavelets is due to Ingrid Daubechies. She formulated the expression for the halfband polynomial P based on the multiresolution analysis conditions in the frequency domain and also in the restriction of finite length filters [18]. Based on this formulation, she discovered the family of extremal phase compactly supported orthogonal wavelets, which are known as Daubechies Wavelets. In her book [20], she gives a good overview of this methodology, and also presents examples of several other families of compactly supported wavelets.

Carl Taswell in [58, 59] presents algorithms for the generation of the Daubechies orthogonal wavelets and for the computation of their regularity.

The book by Gilbert Strang and Truong Nguyen gives a comprehensive treatment of the framework for generating wavelets from a filter bank viewpoint, [55].

Chapter 12
Biorthogonal Wavelets

We have seen in the previous chapters that orthogonality is a very strong constraint for the construction of wavelets. This restricts significantly the design choices of wavelet basis. For example, we showed in Chap. 11 that the Haar wavelet is the only orthogonal basis which is symmetric and has compact support.

A practical solution, that allows more flexibility on the choice of wavelet functions with desirable properties, is to replace orthogonality by a biorthogonality condition.

In this chapter we will introduce biorthogonal wavelet basis, will discuss its relations with perfect reconstruction filter banks, and will present a framework for the design of biorthogonal wavelets.

12.1 Biorthogonal Multiresolution Analysis and Filters

12.1.1 Biorthogonal Basis Functions

Biorthogonal wavelets constitute a generalization of orthogonal wavelets. Under this framework, instead of a single orthogonal basis, a pair of dual biorthogonal basis functions is employed: One for the analysis step and the other for the synthesis step, i.e. we have reciprocal frames as defined in Chap. 2.

J. Gomes, L. Velho, *From Fourier Analysis to Wavelets*,
IMPA Monographs 3, DOI 10.1007/978-3-319-22075-8_12

Recall that, in the context of orthogonal multiresolution analysis, we have defined the projection operators onto the subspaces V_j and W_j, respectively:

$$\text{Proj}_{V_j}(f) = \underbrace{\sum_k \overbrace{\langle f, \phi_{j,k}\rangle}^{\text{analysis}} \phi_{j,k}}_{\text{synthesis}}, \quad \text{and} \quad \text{Proj}_{W_j}(f) = \underbrace{\sum_k \overbrace{\langle f, \psi_{j,k}\rangle}^{\text{analysis}} \psi_{j,k}}_{\text{synthesis}},$$

where the functions ϕ and ψ perform a double duty, i.e. they are used for:

- **analysis**: compute the coefficients of the representation of f in terms of the basis ϕ and ψ of the spaces V_j and W_j, respectively, $c_k^j = \langle f, \phi_{j,k}\rangle$ and $d_k^j = \langle f, \psi_{j,k}\rangle$;
- **synthesis**: reconstruct the projection of f onto V_j and W_j, from the coefficients of the representation, respectively, $\text{Proj}_{V_j}(f) = \sum_k c_k^j \phi_{j,k}$ and $\text{Proj}_{W_j}(f) = \sum_k d_k^j \psi_{j,k}$.

The more general framework of biorthogonal multiresolution analysis employs similar projection operators

$$P_j f = \sum_k \langle f, \phi_{j,k}\rangle \tilde{\phi}_{j,k}, \quad \text{and} \quad Q_j f = \sum_k \langle f, \psi_{j,k}\rangle \tilde{\psi}_{j,k},$$

where the pair of functions ϕ, $\tilde{\phi}$ and ψ, $\tilde{\psi}$ are used to share the workload: one function of the pair acts as the analyzing function, while the other acts as the reconstruction function.

The functions ϕ and ψ are called, respectively, *primal* scaling function and wavelet. The functions $\tilde{\phi}$ and $\tilde{\psi}$ are called, respectively, *dual* scaling function and wavelet. The fact the roles of these functions can be interchanged is called *duality principle.* Although other conventions are possible, here we will assume that the primal functions are used for analysis, while the dual functions are used for synthesis.

In terms of a multiresolution analysis, this scheme leads to a family of *biorthogonal* scaling functions and wavelets that are *dual basis* of the approximating and detail spaces.

More precisely, we define a pair of scaling functions ϕ_j and $\tilde{\phi}_j$ that are, respectively, Riesz basis of the subspaces V_j and $\tilde{V}_j$. Similarly we define a pair of wavelet functions ψ_j and $\tilde{\psi}_j$ that are, respectively, Riesz basis of the subspaces W_j and $\tilde{W}_j$.

These functions generate dual multiresolution analysis ladders

$$\cdots \subset V_1 \subset V_0 \subset V_{-1} \subset \cdots$$
$$\cdots \subset \tilde{V}_1 \subset \tilde{V}_0 \subset \tilde{V}_{-1} \subset \cdots$$

where $V_0 = \text{Span}\{\phi_{0,k}|k \in \mathbb{Z}\}$ and $\tilde{V}_0 = \text{Span}\{\tilde{\phi}_{0,k}|k \in \mathbb{Z}\}$. The spaces W_j and $\tilde{W}_j$ generated by $\psi_{j,k}$ and $\tilde{\psi}_{j,k}$ are, respectively, the complements of V_j in V_{j-1} and of $\tilde{V}_j$ in $\tilde{V}_{j-1}$. In other words, $V_{j-1} = V_j + W_j$ and $\tilde{V}_{j-1} = \tilde{V}_j + \tilde{W}_j$. The intersection of these spaces is null, i.e. $V_j \cap W_j = \{\emptyset\}$ and $\tilde{V}_j \cap \tilde{W}_j = \{\emptyset\}$, but the spaces V_j, W_j, and also $\tilde{V}_j$, $\tilde{W}_j$, are not orthogonal, in general.

In order to compensate for the lack of orthogonality within the approximating and detail spaces, we impose instead a biorthogonality relation between the primal and dual multiresolution ladders, such that

$$V_j \perp \tilde{W}_j \quad \text{and} \quad \tilde{V}_j \perp W_j \tag{12.1}$$

and, consequently,

$$W_j \perp \tilde{W}_l \tag{12.2}$$

for $j \neq l$.

The two multiresolution hierarchies and their sequences of complement spaces fit together according to an intertwining pattern.

The above biorthogonality condition implies that the basis of these spaces must relate as

$$\langle \tilde{\phi}(x), \psi(x-k) \rangle = \int \tilde{\phi}(x)\overline{\psi(x-k)}dx = 0 \tag{12.3}$$

$$\langle \tilde{\psi}(x), \phi(x-k) \rangle = \int \tilde{\psi}(x)\overline{\phi(x-k)}dx = 0 \tag{12.4}$$

and

$$\langle \tilde{\phi}(x), \phi(x-k) \rangle = \int \tilde{\phi}(x)\overline{\phi(x-k)}dx = \delta_k \tag{12.5}$$

$$\langle \tilde{\psi}(x), \psi(x-k) \rangle = \int \tilde{\psi}(x)\overline{\psi(x-k)}dx = \delta_k, \tag{12.6}$$

which can be extended to the multiresolution analysis by a scaling argument resulting in

$$\langle \tilde{\phi}_{j,k}, \phi_{j,m} \rangle = \delta_{k,m}, \qquad j,k,m \in \mathbb{Z} \tag{12.7}$$

$$\langle \tilde{\psi}_{j,k}, \psi_{l,m} \rangle = \delta_{j,l}\delta_{k,m}, \qquad j,k,l,m \in \mathbb{Z}\,. \tag{12.8}$$

12.1.2 Biorthogonality and Filters

The two pairs of scaling function and wavelet, ϕ, ψ, and $\tilde{\phi}$, $\tilde{\psi}$, are defined recursively by the two pairs of filters m_0, m_1, and $\widetilde{m_0}$, $\widetilde{m_1}$.

In the frequency domain these relations are

$$\hat{\phi}(\omega) = m_0(\omega/2)\hat{\phi}(\omega/2), \qquad \hat{\psi}(\omega) = m_1(\omega/2)\hat{\phi}(\omega/2)$$

$$\hat{\tilde{\phi}}(\omega) = \widehat{m_0}(\omega/2)\hat{\tilde{\phi}}(\omega/2), \qquad \hat{\tilde{\psi}}(\omega) = \widehat{m_1}(\omega/2)\hat{\tilde{\phi}}(\omega/2)$$

where

$$m_0(\omega) = \frac{1}{\sqrt{2}}\sum_k h_k\, e^{-ik\omega}, \qquad m_1(\omega) = \frac{1}{\sqrt{2}}\sum_k g_k\, e^{-ik\omega}$$

$$\widetilde{m_0}(\omega) = \frac{1}{\sqrt{2}}\sum_k \tilde{h}_k\, e^{-ik\omega}, \qquad \widetilde{m_1}(\omega) = \frac{1}{\sqrt{2}}\sum_k \tilde{g}_k\, e^{-ik\omega}\ .$$

By computing the Fourier Transform of the inner products in Eqs. (12.3) to (12.6), and using the same argument of Chap. 10 for the characterization of m_0 and m_1, we can see that the biorthogonality condition in the frequency domain is equivalent to

$$\sum_k \hat{\tilde{\phi}}(\omega + k2\pi)\overline{\hat{\phi}(\omega + k2\pi)} = 1$$

$$\sum_k \hat{\tilde{\psi}}(\omega + k2\pi)\overline{\hat{\psi}(\omega + k2\pi)} = 1$$

$$\sum_k \hat{\tilde{\psi}}(\omega + k2\pi)\overline{\hat{\phi}(\omega + k2\pi)} = 0$$

$$\sum_k \hat{\tilde{\phi}}(\omega + k2\pi)\overline{\hat{\psi}(\omega + k2\pi)} = 0$$

for all $\omega \in \mathbb{R}$.

This means that the filters m_0, m_1 and their duals $\widetilde{m_0}$, $\widetilde{m_1}$ have to satisfy

$$\widetilde{m_0}(\omega)\overline{m_0(\omega)} + \widetilde{m_0}(\omega+\pi)\overline{m_0(\omega+\pi)} = 1 \tag{12.9}$$

$$\widetilde{m_1}(\omega)\overline{m_1(\omega)} + \widetilde{m_1}(\omega+\pi)\overline{m_1(\omega+\pi)} = 1 \tag{12.10}$$

$$\widetilde{m_1}(\omega)\overline{m_0(\omega)} + \widetilde{m_1}(\omega+\pi)\overline{m_0(\omega+\pi)} = 0 \tag{12.11}$$

$$\widetilde{m_0}(\omega)\overline{m_1(\omega)} + \widetilde{m_0}(\omega+\pi)\overline{m_1(\omega+\pi)} = 0\ . \tag{12.12}$$

The set of equations above can be written in matrix form as

$$\begin{pmatrix} \widetilde{m_0}(\omega) & \widetilde{m_0}(\omega+\pi) \\ \widetilde{m_1}(\omega) & \widetilde{m_1}(\omega+\pi) \end{pmatrix} \overline{\begin{pmatrix} m_0(\omega) & m_1(\omega) \\ m_0(\omega+\pi) & m_1(\omega+\pi) \end{pmatrix}} = \begin{pmatrix} 1 & 0 \\ 0 & 1 \end{pmatrix}$$

or

$$\tilde{M}(\omega)\overline{M^T(\omega)} = I \tag{12.13}$$

where M is the modulation matrix introduced in Chap. 10

$$M(\omega) = \begin{bmatrix} m_0(\omega) & m_0(\omega+\pi) \\ m_1(\omega) & m_1(\omega+\pi) \end{bmatrix} \tag{12.14}$$

and I is the identity matrix.

12.1.3 Fast Biorthogonal Wavelet Transform

Because ϕ and ψ define a multiresolution analysis, we have that

$$\phi(x) = \sum_k h_k \phi(2x-k) \quad \text{and} \quad \psi(x) = \sum_k g_k \phi(2x-k) \, . \tag{12.15}$$

Similarly, $\tilde{\phi}$ and $\tilde{\psi}$ also define a multiresolution analysis, and therefore

$$\tilde{\phi}(x) = \sum_k \tilde{h}_k \tilde{\phi}(2x-k) \quad \text{and} \quad \tilde{\psi}(x) = \sum_k \tilde{g}_k \tilde{\phi}(2x-k) \, . \tag{12.16}$$

We can derive the coefficients of the filters $\tilde{m}_0$ and $\tilde{m}_1$ by combining the above two equations with the biorthogonality relations $\langle \tilde{\phi}_{j,k}, \phi_{j,m} \rangle = \delta_{k,m}$ and $\langle \tilde{\psi}_{j,k}, \psi_{l,m} \rangle = \delta_{j,l}\delta_{k,m}$. This gives us

$$\tilde{h}_{k-2l} = \langle \tilde{\phi}(x-l), \phi(2x-k) \rangle \tag{12.17}$$

$$\tilde{g}_{k-2l} = \langle \tilde{\psi}(x-l), \phi(2x-k) \rangle \tag{12.18}$$

and also

$$h_{k-2l} = \langle \phi(x-l), \tilde{\phi}(2x-k) \rangle \tag{12.19}$$

$$g_{k-2l} = \langle \psi(x-l), \tilde{\phi}(2x-k) \rangle \, . \tag{12.20}$$

By writing $\phi(2x-k) \in V_{-1}$ in terms of the bases of V_0 and W_0, we get the two-scale relation

$$\phi(2x-k) = \sum_l \tilde{h}_{k-2l}\phi(x-l) + \sum_l \tilde{g}_{k-2l}\psi(x-l) \tag{12.21}$$

and, since primary and dual functions are interchangeable, we also have

$$\tilde{\phi}(2x-k) = \sum_l h_{k-2l}\tilde{\phi}(x-l) + \sum_l g_{k-2l}\tilde{\psi}(x-l) \ . \tag{12.22}$$

The fast biorthogonal wavelet transform uses the above decomposition/reconstruction relation. The algorithm employs the two pairs of primary and dual filters and except for this difference, it is essentially similar to the orthogonal case, presented in Chap. 7.

The pair of filters $\widetilde{m_0}$, $\widetilde{m_1}$ is employed in the decomposition step and the pair of filters m_0, m_1 in the reconstruction step.

In the decomposition step we employ a discrete convolution with the filter coefficients (h_k) and (g_k) of the filters m_0 and m_1.

$$c_n^{j+1} = \sum_k h_{k-2n} c_k^j \tag{12.23}$$

and

$$d_n^{j+1} = \sum_k g_{k-2n} c_k^j \ . \tag{12.24}$$

Conversely, in the reconstruction step, we employ a discrete convolution with the filter coefficients, $(\tilde{h}_k)$ and $(\tilde{g}_k)$, of the filters $\widetilde{m_0}$ and $\widetilde{m_1}$

$$c_l^{j-1} = \sum_n \tilde{h}_{l-2n} c_n^j + \sum_n \tilde{g}_{l-2n} c_n^j \ . \tag{12.25}$$

Remember that, as we already noted, the roles of these two filter banks can be interchanged.

12.2 Filter Design Framework for Biorthogonal Wavelets

12.2.1 Perfect Reconstruction Filter Banks

The filters m_0, $\widetilde{m_0}$, m_1, $\widetilde{m_1}$ define a two-channel filter bank, where m_0, m_1 are, respectively, the low-pass and high-pass filters used in the analysis step, and $\widetilde{m_0}$, $\widetilde{m_1}$ are, respectively, the low-pass and high-pass filters used in the synthesis step, as shown in Fig. 12.1. Once again, we note that the role of these filters can be interchanged.

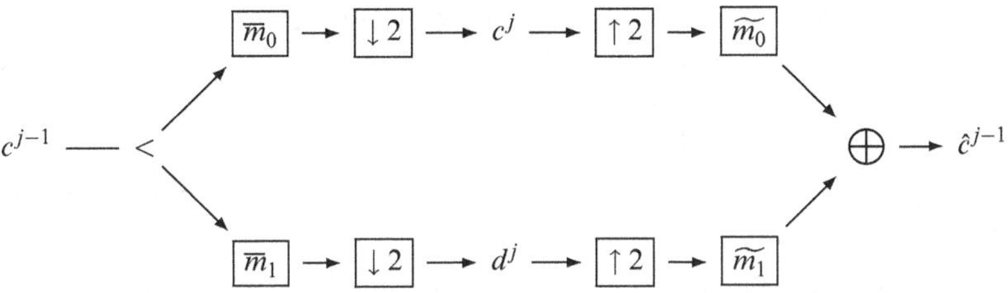

Fig. 12.1 Biorthogonal Filter Bank

We would like this two-channel filter bank to have the property of *perfect reconstruction*. This means that a signal can be exactly reconstructed by the synthesis filters from the coefficients of the representation constructed by the analysis filters. These conditions can be derived by following a discrete signal, (x_n), through the filter bank.

The output of the low-pass channel can be written in a compact notation using the z-notation (see Appendix B).

We will do this in three stages: First, we perform a discrete convolution of analysis low-pass filter with the signal, whose z-transform is $\overline{m}_0(z)x(z)$. This is followed by downsampling and upsampling, which in the z-domain is $1/2\ [\overline{m}_0(z)x(z) + \overline{m}_0(-z)x(-z)]$. Finally, we perform a discrete convolution of the resulting coefficient sequence with the synthesis filter, whose z-transform is

$$\frac{1}{2}\,\widetilde{m_0}(z)\,[\overline{m}_0(z)x(z) + \overline{m}_0(-z)x(-z)]\ . \tag{12.26}$$

The output of the high-pass channel is obtained in a similar manner

$$\frac{1}{2}\,\widetilde{m_1}(z)\,[\overline{m}_1(z)x(z) + \overline{m}_1(-z)x(-z)]\ . \tag{12.27}$$

The filter bank combines the outputs of the low-pass and high-pass channel by adding these two expressions, which gives

$$\begin{aligned}\hat{x}(z) &= \frac{1}{2}\ [\widetilde{m_0}(z)\overline{m}_0(z) + \widetilde{m_1}(z)\overline{m}_1(z)]\ x(z)\\ &+ \frac{1}{2}\ [\widetilde{m_0}(z)\overline{m}_0(-z) + \widetilde{m_1}(z)\overline{m}_1(-z)]\ x(-z)\end{aligned}$$

where we rearranged the expression to separate the term involving $x(z)$ from the terms involving $x(-z)$.

Note that the downsampling/upsampling operators introduced *aliasing* in the signal, manifested by the appearance of terms $x(-z)$, as well as $x(z)$. By substituting, $z = e^{-i\omega}$ and $-z = e^{-i\omega+\pi}$, it is apparent that the frequencies $w+\pi$ appear as aliases of frequencies w in the signal (see Appendix A for details).

For perfect reconstruction we must have $\hat{x}(z) = x(z)$, which implies in two conditions:

- **alias cancellation**: the aliasing components caused by subsampling in the low-pass channel and in the high-pass channel cancel each other.

$$\widetilde{m_0}(z)\overline{m}_0(-z) + \widetilde{m_1}(z)\overline{m}_1(-z) = 0$$

or, making $z = e^{-i\omega}$

$$\widetilde{m_0}(\omega)\overline{m_0(\omega+\pi)} + \widetilde{m_1}(\omega)\overline{m_1(\omega+\pi)} = 0 \tag{12.28}$$

- **no distortion**: the signal is reconstructed without loss or gain of energy.

$$\widetilde{m_0}(z)\overline{m}_0(z) + \widetilde{m_1}(z)\overline{m}_1(z) = 2$$

or

$$\widetilde{m_0}(\omega)\overline{m_0(\omega)} + \widetilde{m_1}(\omega)\overline{m_1(\omega)} = 2\,. \tag{12.29}$$

12.2.2 Conjugate Quadrature Filters

We have four filters to design, m_0, m_1, $\widetilde{m_0}$, $\widetilde{m_1}$. These filters must satisfy the conditions for perfect reconstruction. An interesting option is to investigate how we can determine some of the filters from the others, such that conditions (12.28) and (12.29) are automatically satisfied.

First, we use a conjugate quadrature scheme as done in Chap. 10, to define the high-pass filters in terms of the low-pass filters:

$$m_1(\omega) = e^{-i\omega}\overline{\widetilde{m_0}(\omega+\pi)} \tag{12.30}$$

$$\widetilde{m_1}(\omega) = e^{-i\omega}\overline{m_0(\omega+\pi)}\,. \tag{12.31}$$

This option takes care of the alias cancelation, and guarantees that Eq. (12.28) is satisfied, as it can seen by substituting (12.30), (12.31) into (12.28).

Next, to express the no-distortion condition in terms of only m_0 and $\widetilde{m_0}$, we rewrite Eq. (12.29) using (12.30), (12.31).

$$\widetilde{m_0}(\omega)\overline{m_0(\omega)} + \widetilde{m_0}(\omega+\pi)\overline{m_0(\omega+\pi)} = 2 \tag{12.32}$$

Note that, when we normalize the filters (multiplying their coefficients by $1/\sqrt{2}$), Eq. (12.32) is exactly (12.9), the first condition for biorthogonality of the filters m_0, $\widetilde{m_0}$, m_1, $\widetilde{m_1}$. The other conditions (12.10), (12.12), and (12.11) can be derived from the perfect reconstruction conditions (12.28) and (12.29), using the conjugate quadrature relations (12.30) and (12.31). This demonstrates that biorthogonal wavelets are associated with a subband filtering scheme with exact reconstruction.

12.2.3 The Polynomial P and Wavelet Design

Now, we define the product filter $P(\omega) = \widetilde{m_0}(\omega)\overline{m_0(\omega)}$, and Eq. (12.32) becomes

$$P(\omega) + P(\omega + \pi) = 2 \,. \tag{12.33}$$

To design the filters m_0, $\widetilde{m_0}$, m_1, $\widetilde{m_1}$ that generate biorthogonal wavelets, we use the following procedure:

1. Choose a polynomial P satisfying Eq. (12.33)
2. Factor P into $\widetilde{m_0}\ \overline{m_0}$
3. Use (12.30) and (12.31) to obtain m_1 and $\widetilde{m_1}$.

Observe that this procedure is very similar to the one described in Chap. 10. In fact, we are using the same filter design framework to create wavelet basis. The main difference is that now, the analysis filters can be different from the synthesis filters, because we have two biorthogonal multiresolution hierarchies. This gives us more freedom to design wavelets with desired properties.

As we have seen in Chap. 10, in order to satisfy the perfect reconstruction condition, the polynomial $P(\omega) = \sum_{k=-N}^{N} = a_k e^{ik\omega}$ must be a halfband filter, i.e. the terms with index k even must have coefficient $a_k = 0$, except for $a_0 = 1$, such that the terms with index k odd, which have coefficients $a_k \neq 0$, will cancel in Eq. (12.33), because $e^{i2k(\omega+\pi)} = -e^{i2k\omega}$. The only remaining term will be $a_0 = 1$, resulting in $P(\omega) + P(\omega + \pi) = 2$ (or when the filters are normalized $a_0 = 1/2$ and $P(\omega) + P(\omega + \pi) = 1$).

Also, because both ϕ and $\tilde{\phi}$ define multiresolution analyses, the polynomials $m_0(\omega)$ and $\widetilde{m_0}(\omega)$ must have at least one zero at π (see Chap. 11 for details). Therefore, $P(\omega) = \widetilde{m_0}(\omega)\overline{m_0(\omega)}$ has to be of the form

$$P(\omega) = \left|\left(1 + e^{i\omega}\right)^n\right|^2 B(e^{i\omega}) \,. \tag{12.34}$$

The advantage of biorthogonal wavelets over orthogonal wavelets is that we can factorize $P(\omega)$ in two different polynomials m_0, $\widetilde{m_0}$ instead as the "square" of a single polynomial m_0.

12.2.4 Factorization of P for Biorthogonal Filters

Let us see one example of the different design choices for the factorization of $P(\omega)$ into $m_0(\omega)$ and $\widetilde{m_0}(\omega)$, from [55].

We will design the polynomial P such that it will have four zeros at π

$$\begin{aligned} P(\omega) &= |(1+e^{-i\omega})^2|^2 B(e^{i\omega}) \\ &= (1+e^{-i\omega})^2(1+e^{i\omega})^2 B(e^{-i\omega}) \,. \end{aligned}$$

Making $z = e^{-i\omega}$

$$\begin{aligned} P(z) &= (1+z^{-1})^2(1+z)^2 B(z) \\ &= (z^{-2}+4z^{-1}+6+4z+z^2)B(z) \\ &= z^2(1+z^{-1})^4 \, B(z) \,. \end{aligned}$$

We have to choose B so that even powers are eliminated from P, making it a halfband polynomial. The solution is $B(z) = -z^{-1}+4-z$

$$\begin{aligned} P(z) &= (z^{-2}+4z^{-1}+6+4z+z^2)(-z^{-1}+4-z) \\ &= -z^{-3}+9z^{-1}+16+9z-z^3 \,. \end{aligned}$$

We normalize it to make $a_0 = 1$

$$P(z) = \frac{1}{16}(-z^{-3}+9z^{-1}+16+9z-z^3) \,. \tag{12.35}$$

Note that P satisfies the perfect reconstruction conditions and also has four zeros at -1.

We have several choices to factorize $P(z)$ into $\widetilde{m_0}(z)\overline{m_0(z)}$. The polynomial P has six roots $\{-1,-1,-1,-1,2-\sqrt{3},2+\sqrt{3}\}$. The four roots at $z=-1$ come from $(1+z^{-1})^4$, and the other two roots at $c = 2-\sqrt{3}$ and $1/c = 2+\sqrt{3}$ come from $B(z)$.

Each of the polynomials $m_0(z)$ and $\widetilde{m_0}(z)$ will have, at least, one root at $z=-1$. Therefore, we can select the roots for the filter m_0 (or $\widetilde{m_0}$) in one of the following ways, — the other filter $\widetilde{m_0}$ (or m_0) will have the remaining roots:

i. m_0 of degree 1: $(1+z^{-1})$
ii. m_0 of degree 2: $(1+z^{-1})^2$, or $(1+z^{-1})\,(c-z^{-1})$, or $(1+z^{-1})\,(1/c-z^{-1})$
iii. m_0 of degree 3: $(1+z^{-1})^3$, or $(1+z^{-1})^2\,(c-z^{-1})$.

We have six different choices for the factorization of P. For each choice the roots are separated in two sets, and we can select which set of roots go to the analysis filter m_0 and to the synthesis filter $\widetilde{m_0}$.

In option (i), one factor is $(1 + z^{-1})$, which is the filter for a box function. The other factor is a degree 5 polynomial $\frac{1}{16}(-1 + z^{-1} + 8z^{-2} + 8z^{-3} + z^{-4} - z^{-5})$. The filter bank will have filters with 2 and 6 coefficients for analysis and synthesis, respectively.

In option (ii), the first possibility is one factor equal to $\frac{1}{2}(1 + 2z^{-1} + z^{-2})$, which is the filter for the hat function, and the other factor equal to $\frac{1}{8}(-1 + 2z^{-1} + 6z^{-2} + 2z^{-3} - z^{-4})$. This choice corresponds to the order 2 B-spline biorthogonal wavelet, that will be discussed in the next section. It is best to use the hat scaling function for synthesis, resulting in a 5/3 filter bank. The other possibilities are not so interesting.

In option (iii), we have two different possibilities, both very important. The first possibility is a factorization as $\frac{1}{8}(1+3z^{-1}+3z^{-2}+z^{-3})$, and $\frac{1}{2}(-1+3z^{-1}+3z^{-2}-z^{-3})$. Notice that this choice results in a pair of linear phase filters which correspond to symmetric scaling functions and anti-symmetric wavelets. The second possibility is a factorization with $m_0(z) = \widetilde{m_0}(z) = \frac{1}{4\sqrt{2}}[(1 + \sqrt{3}) + (3 + \sqrt{3})z^{-1} + (3 - \sqrt{3})z^{-2} + (1 - \sqrt{3})z^{-3}]$. This choice gives us the Daubechies orthogonal wavelet of order 2, that we presented in the previous chapter. Notice that the orthogonal filters have extremal phase, which correspond to the most asymmetric scaling functions and wavelets.

Figure 12.2 shows the graph of the biorthogonal scaling functions and wavelets derived in case (i), and Fig. 12.3 plots the frequency response of the corresponding filters.

Figure 12.4 shows the graph of the biorthogonal scaling functions and wavelets derived in case (ii), and Fig. 12.5 plots the frequency response of the corresponding filters.

Figure 12.6 shows the graph of the biorthogonal scaling functions and wavelets derived in option (a) of case (iii), and Fig. 12.7 plots the frequency response of the corresponding filters.

Figure 12.8 shows the graph of the orthogonal D2 Daubechies scaling functions and wavelets derived in option (b) of case (iii), and Fig. 12.9 plots the frequency response of the corresponding filters.

12.3 Symmetric Biorthogonal Wavelets

One of the advantages of biorthogonal over orthogonal wavelets is that we can have symmetry. In this section we present some families of symmetric biorthogonal wavelets, developed by Ingrid Daubechies [20], using different factorizations of the polynomial P.

For symmetric filters we have to factorize P such that the roots z_i and z_i^{-1} must stay together when we split the roots between m_0 and $\widetilde{m_0}$. Therefore, we have

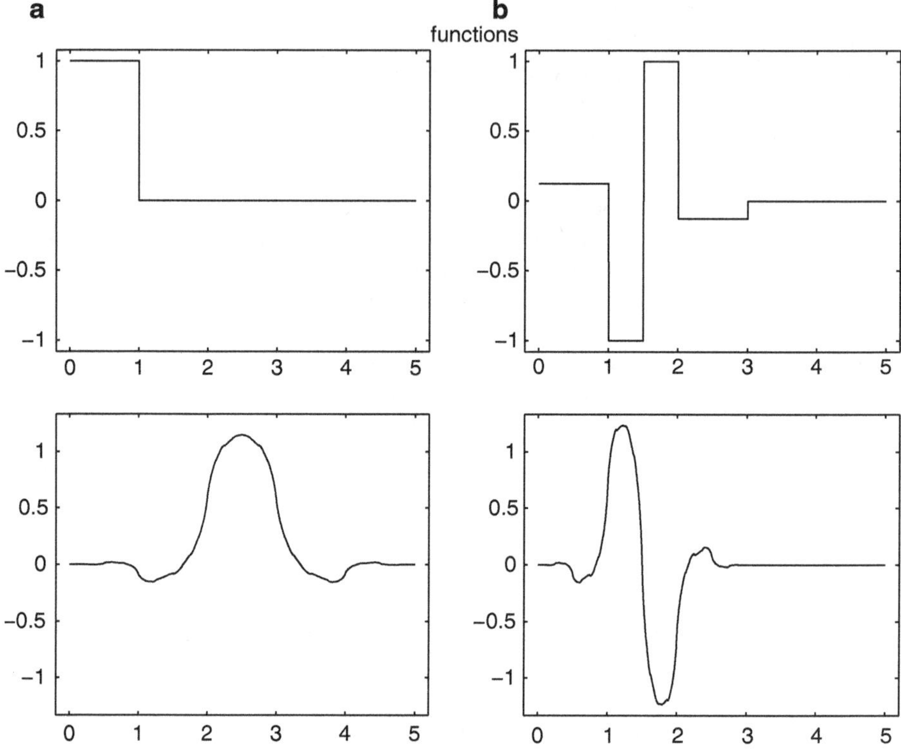

Fig. 12.2 Biorthogonal box scaling functions **a** and wavelets **b**

$$z^N P(z) = a^N \prod_{i=1}^{N} (z - z_i) \left(z - \frac{1}{z_i} \right) \tag{12.36}$$

where each pair of roots, $\{z_i, 1/z_i\}$, must be assigned either to m_0 or to $\widetilde{m_0}$.

12.3.1 B-Spline Wavelets

B-Splines of order N are piecewise polynomials that come from the convolution of N box functions. In terms of the two-scale relations, they can be constructed from a filter function $m_0(z)$ of the form

$$m_0(z) = \left(\frac{1 + z^{-1}}{2} \right)^N . \tag{12.37}$$

Each factor, $(\frac{1+z^{-1}}{2})$, corresponds to a convolution with a box function.

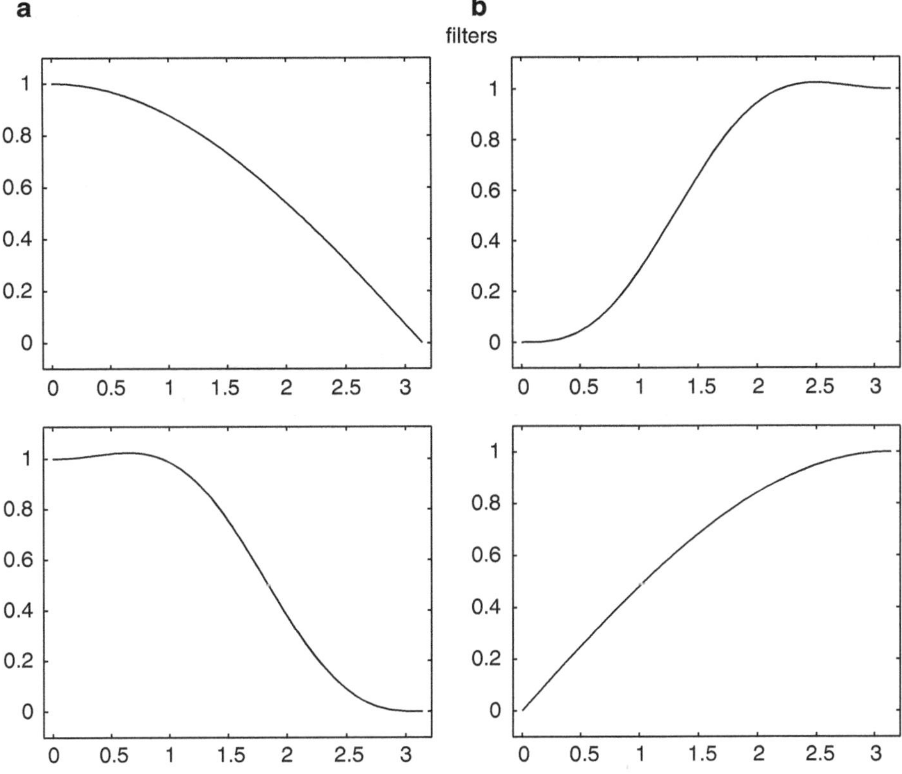

Fig. 12.3 Frequency response of low-pass **a** and high-pass **b** filters

To construct biorthogonal B-spline wavelets we can split the polynomial P introduced in the previous chapter as:

$$\widetilde{m_0}(z) = \left(\frac{1+z^{-1}}{2}\right)^{\tilde{M}} \tag{12.38}$$

where $\tilde{M} = 2\tilde{\ell}$ is even or $\widetilde{M} = 2\tilde{\ell} + 1$ is odd, and

$$m_0(z) = \left(\frac{1+z^{-1}}{2}\right)^{M} \sum_{m=0}^{\ell+\tilde{\ell}-\sigma} \binom{\ell+\tilde{\ell}-\sigma+m}{m} \left(\frac{1-z}{2}\right)^{m} \left(\frac{1-z^{-1}}{2}\right)^{m} \tag{12.39}$$

where $M = 2\ell$ is even, or $M = 2\ell + 1$ is odd, and $\sigma = 0$ if M is odd, or $\sigma = 1$ if M is even.

Note that this formulas give explicit expressions for $\widetilde{m_0}$ and m_0. We have, thus a family of biorthogonal B-spline wavelets, in which the synthesis scaling function $\tilde{\phi}$ is a B-spline of order $\tilde{M}$. For a fixed $\tilde{M}$, there is an infinity of choices for M,

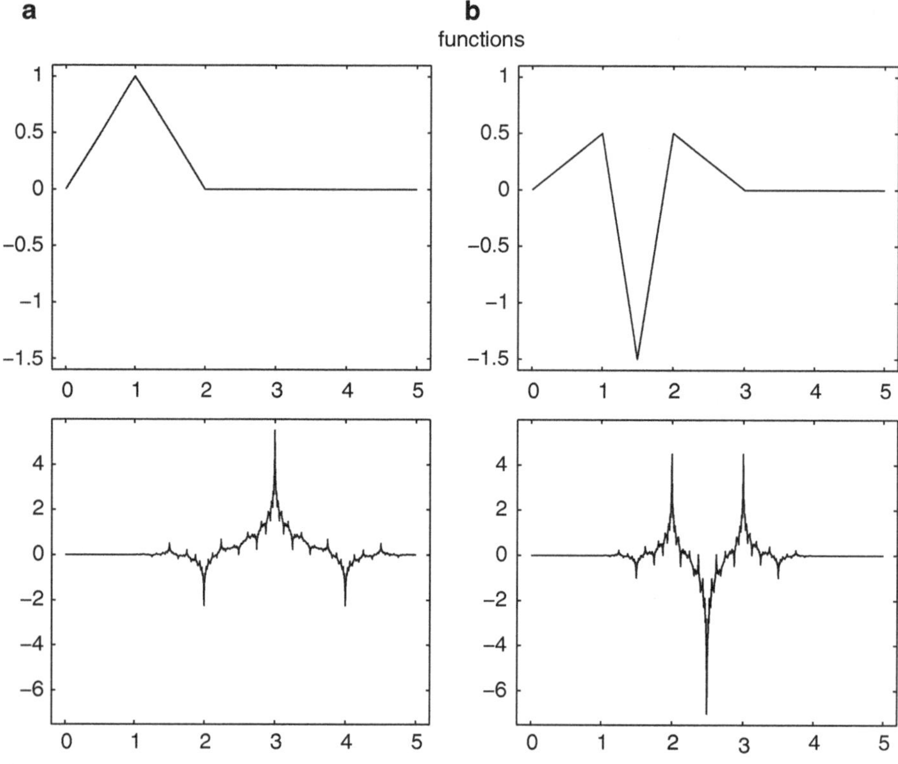

Fig. 12.4 Biorthogonal Hat scaling functions **a** and wavelets **b**

resulting in biorthogonal functions with increasing regularity and support width. Also, we remark that $\widetilde{m_0}$ depends only on $\tilde{M}$, while m_0 depends on both M and $\tilde{M}$.

In Fig. 12.10 we show one example of a biorthogonal B-spline wavelets and scaling functions of order 2 (linear spline), $\tilde{M} = 2$, with $M = 2$ and 4.

This family of wavelets has two main advantages: First, there is a closed form expression for the B-spline scaling functions $\tilde{\phi}$ and wavelets $\tilde{\psi}$; Second, the coefficients for all filters $\widetilde{m_0}$, $\widetilde{m_1}$, m_0, m_1 are dyadic rationals, which make them suitable for fast and exact computer implementations.

The major disadvantage of biorthogonal B-spline wavelets is that the analysis and synthesis functions have very different support widths, as can be seen in Fig. 12.10. This may or may not be a problem in some applications.

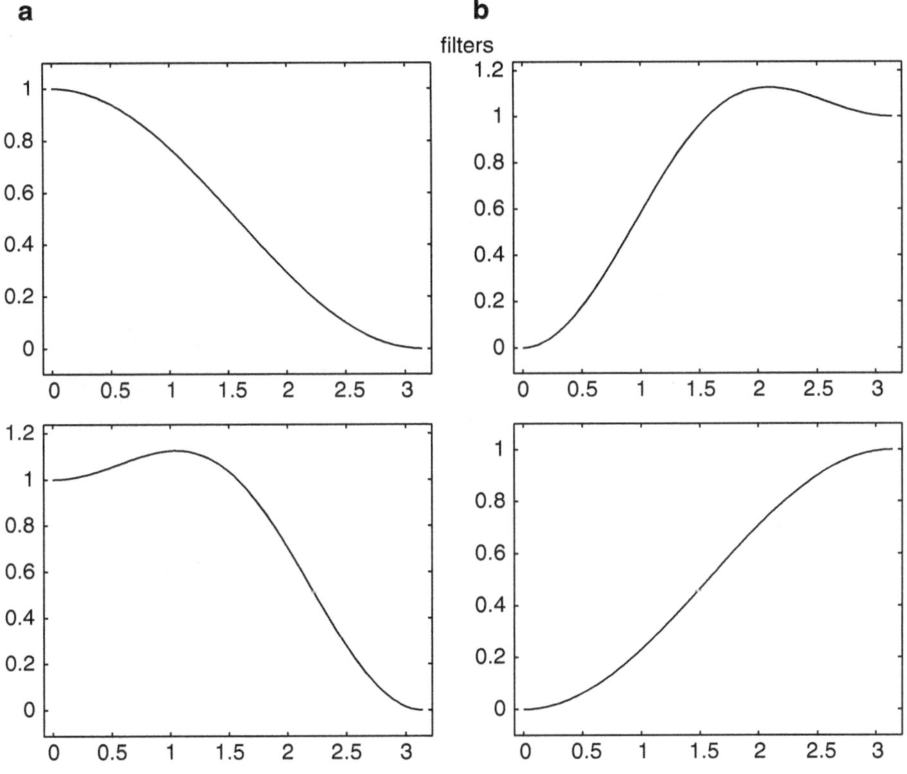

Fig. 12.5 Frequency response of low-pass **a** and high-pass **b** filters

12.3.2 Wavelets with Closer Support Width

It is possible to construct wavelets with closer support width by choosing an appropriate factorization of the polynomial P. The goal is to find filter functions m_0 and $\widetilde{m_0}$, that have both linear phase and similar degree.

We determine all roots of P, real zeros r_i, and pairs of complex conjugate zeros $\{z_j, \overline{z}_j\}$

$$P(z) = c \prod_{i=1} (z - r_i) \quad \prod_{j=1} (z - z_j)(z - \overline{z}_j) \,. \tag{12.40}$$

For a fixed degree $N = \ell + \tilde{\ell}$, we have a limited number of factorizations. Therefore, we can generate all the different combinations of roots, such that r_i, $\{z_j, \overline{z}_j\}$ are assigned either to m_0 or to $\widetilde{m_0}$, and select the option $\ell + \tilde{\ell}$, that makes the lengths of m_0 and $\widetilde{m_0}$ as close as possible.

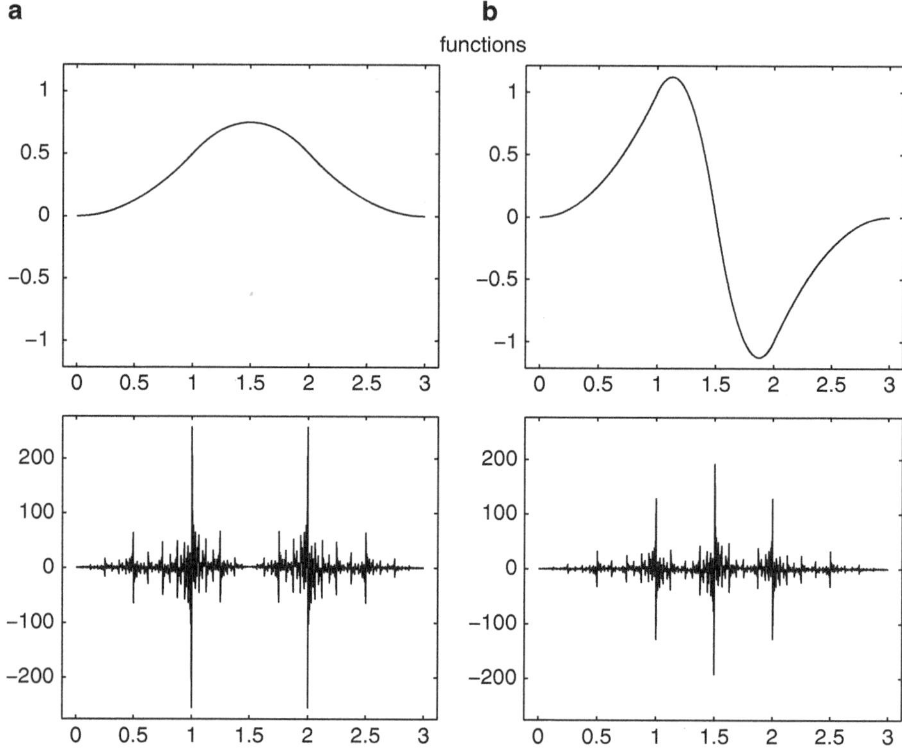

Fig. 12.6 Biorthogonal quadratic B-spline scaling functions **a** and wavelets **b**

Figure 12.11 shows an example of symmetric biorthogonal wavelets designed using this method.

12.3.3 Biorthogonal Bases Closer to Orthogonal Bases

Motivated by a suggestion of M. Barlaud, Ingrid Daubechies developed a family of biorthogonal wavelet bases that are close to an orthogonal basis. Barlaud tried to construct biorthogonal wavelets using the Laplacian pyramid filter, designed by P. Burt [8], as either m_0 or $\widetilde{m_0}$. These experiments lead to the discovery that the Burt filter is very close to an orthonormal wavelet filter.

The particular construction for the Burt filter was then generalized by Daubechies to a family of biorthogonal wavelets that are close to orthogonal. She used the extended formula for the polynomial P, including the factor $R(z^N)$, as discussed in the previous chapter.

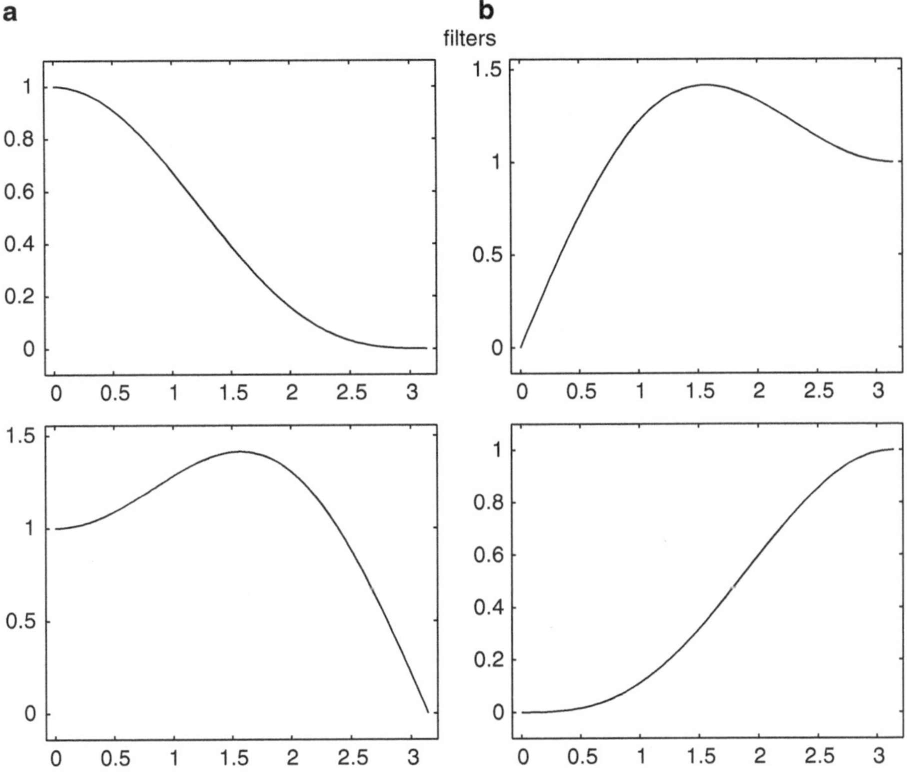

Fig. 12.7 Frequency response of low-pass **a** and high-pass **b** filters

The filter functions for this family are defined as

$$m_0(z) = S(z) + aR(z), \quad \text{and} \quad \widetilde{m_0}(z) = S(z) + bR(z) \tag{12.41}$$

where the S, R are given below, and the constants a, b are computed by an optimization procedure to guarantee the biorthogonality of the filters.

$$S(z) = \left(\frac{1+z}{2}\right)^{2K} \left(\frac{1+z^{-1}}{2}\right)^{2K} \sum_{k=0}^{K-1} \binom{K-1+k}{k} \left(\frac{1-z}{2}\right)^{2k} \left(\frac{1-z^{-1}}{2}\right)^{2k} \tag{12.42}$$

$$R(z) = \left(\frac{1-z^{-2}}{4}\right)^{2K} \left(\frac{1-z^{2}}{4}\right)^{2K} \tag{12.43}$$

Note that the filters m_0 and $\widetilde{m_0}$ differ only by the constants a and b, therefore they are very close to an orthogonal filter.

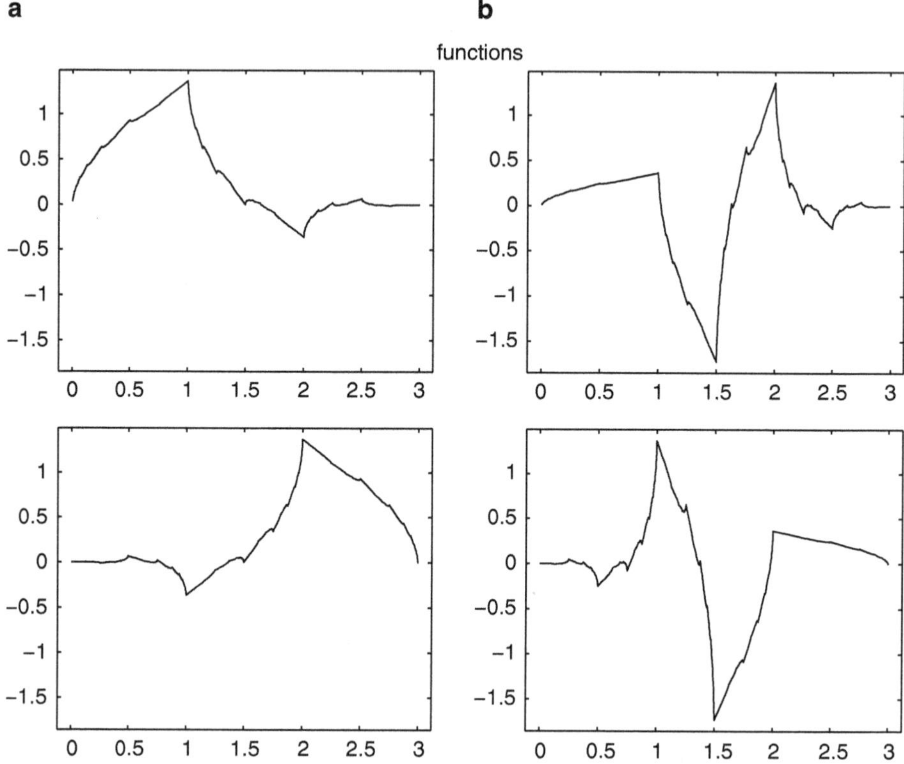

Fig. 12.8 D2 Daubechies scaling functions **a** and wavelets **b**

Figure 12.12 shows a plot of a wavelet constructed using this procedure.

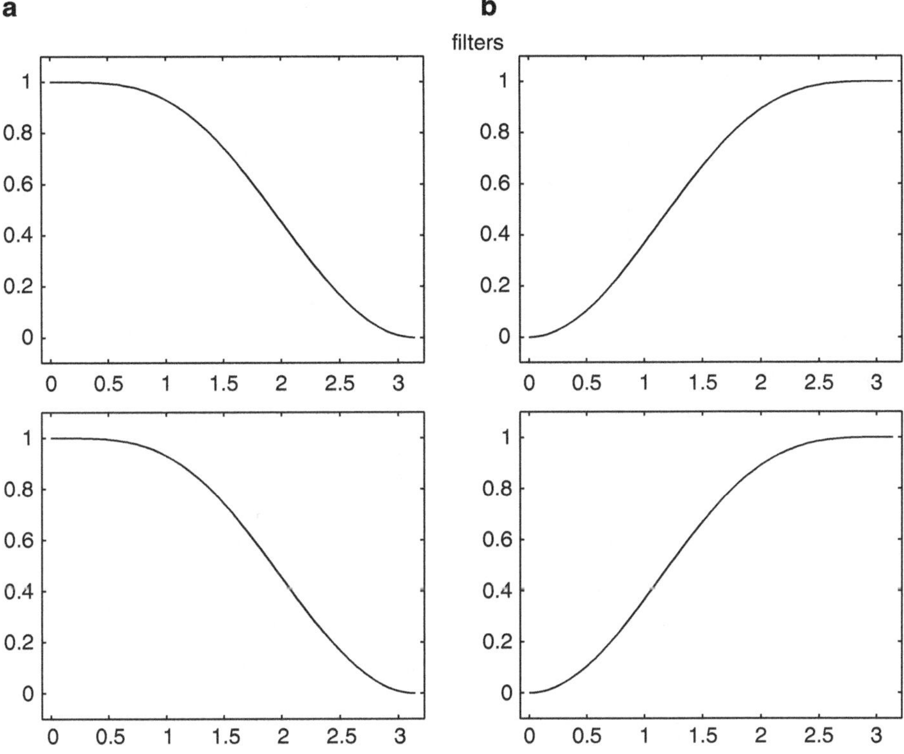

Fig. 12.9 Frequency response of low-pass **a** and high-pass **b** filters

12.4 Comments and References

The first examples of biorthogonal wavelets were developed independently by [14] and [62].

It is possible to construct biorthogonal wavelets such that the primal and dual scaling and wavelet bases generate a single orthogonal multiresolution analysis. In this case, the scaling function and wavelets are called *semi-orthogonal*. The name "pre-wavelets" is also employed to designate this type of wavelets. An extensive treatment of this subject can be found in [13].

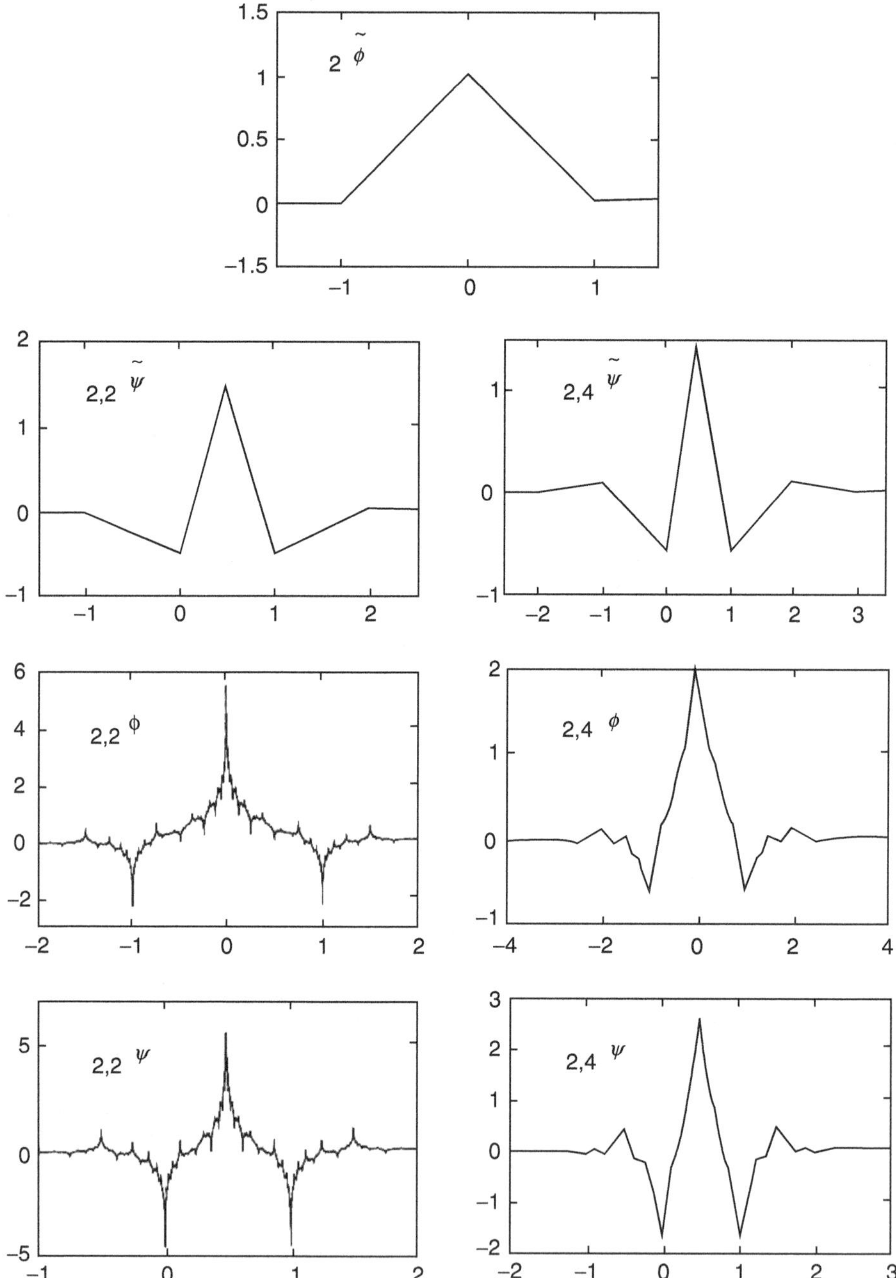

Fig. 12.10 Biorthogonal B-spline Wavelets of order 2 (from [20])

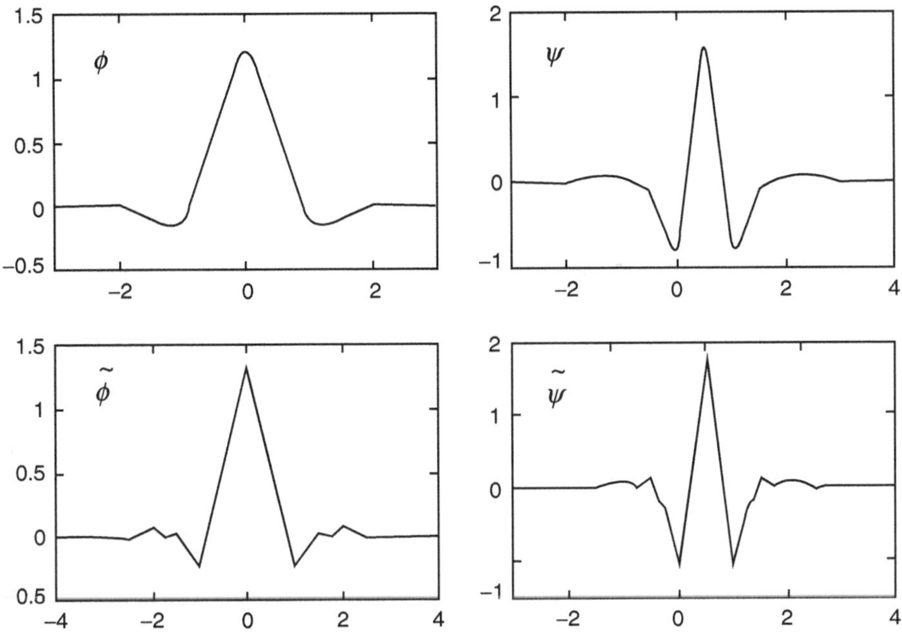

Fig. 12.11 Biorthogonal Wavelets that have Similar Support, (from [20])

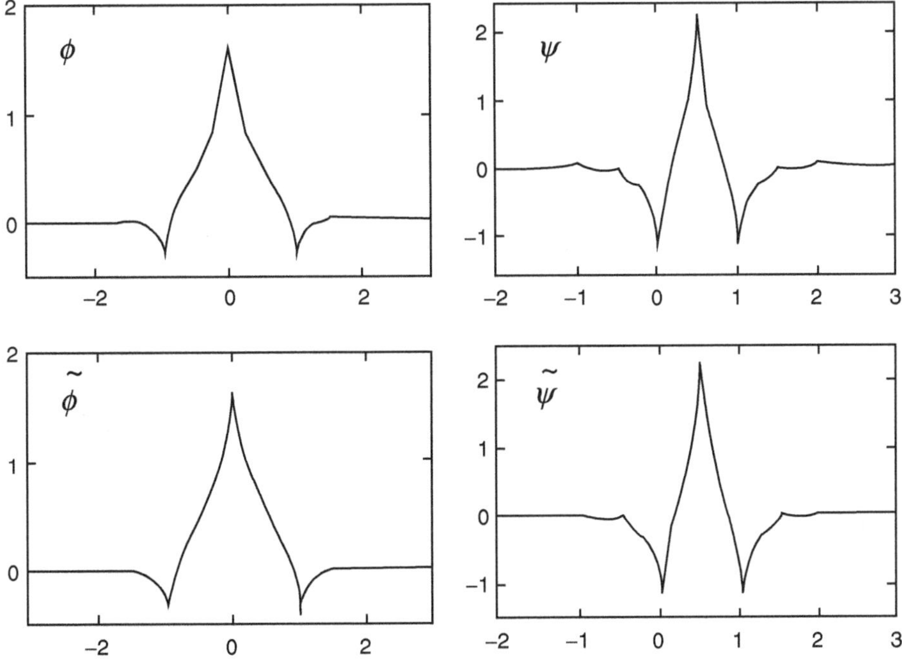

Fig. 12.12 Biorthogonal Wavelets Close to Orthogonal (from [20])

Chapter 13
Directions and Guidelines

In this concluding chapter we will review the material covered in the book, and give some directions for further studies on the subject of wavelets.

13.1 History and Motivation

The field of wavelets, despite being relatively recent, is vast and is developing very rapidly. This is true in relation to both the theoretical aspects of wavelets, and the applications.

Wavelets are a product of groundwork from many areas, ranging from pure mathematics and physics to engineering and signal processing. Independent research in these areas pursued similar goals using different approaches. The objective was to develop tools to describe functions in time *and* frequency simultaneously. The separate lines of investigation reached a mature point, and in the beginning of the 1980s, the confluence of this interdisciplinary sources was formalized originating the theory of wavelets. The subsequent unification of the field was a key factor to make wavelets popular in applied mathematics, and also to give a significant impulse to new research.

Today, wavelets are well established. The basic theory is completely developed with successful applications in a large number of areas. Nonetheless, research is perhaps even more active than before, with new results appearing from a growing scientific community. Also, the application base is consolidating and expanding to new areas, with new experimental systems and commercial products being released.

Considering the facts mentioned above, it would not be possible to cover the whole field of wavelets in a single text. Indeed, there are currently more than 20 textbooks, entirely dedicated to the various aspects of wavelets. Not to mention, an enormous number of conference proceedings and special issues of journals devoted to wavelets. Most of these texts are for experts, or for people with basic knowledge

J. Gomes, L. Velho, *From Fourier Analysis to Wavelets*,
IMPA Monographs 3, DOI 10.1007/978-3-319-22075-8_13

of wavelets. Even some of the introductory books cover a lot of material with a fast pace, assuming maturity and dedication from the reader. For this reason, we believe that there is still a need for a conceptual textbook about the fundamentals of wavelets.

13.2 A Look Back

In this book, we tried to give a comprehensive, but accessible, introduction to the field of wavelets. Our goal was to follow a route from Fourier Analysis to wavelets, showing how the traditional tools for function representation in frequency domain evolved to the new tools for joint time-frequency descriptions. This approach combined the concepts from function analysis with the intuition from signal processing.

It is instrumental to take a look back and review the main topics that we covered in this book. They include:

- motivation and schemes for representation of functions;
- the Fourier transform as a tool for frequency analysis;
- the Windowed Fourier transform and the search for time-frequency localization;
- the continuous Wavelet transform as an adapted time-frequency function decomposition;
- the multiresolution representation as a tool to discretize and reconstruct functions;
- discrete wavelet bases and a filter bank methodology for design and computation;
- and finally we described the main families of wavelet bases.

As we stated earlier, these items constitute only the fundamental concepts behind wavelets. In order to go from this basic level to more advanced topics and to practical applications, the reader can continue the learning process in various directions. Below, we will discuss some of the important aspects of wavelets not covered in this book, and will indicate some of the possible options for further studies.

In this book, we restricted ourselves to functions defined on the whole real line, which correspond to signals with infinite duration. This is the simplest case, and certainly makes the theory more accessible.

We also discussed only the basic schemes for time-frequency decomposition of a function. In particular, we have emphasized descriptions using non-redundant wavelet bases. This is perhaps the most important representation in practical applications.

Lastly, we didn't consider any concrete application of wavelet, besides a few simple example scattered throughout the book.

13.3 Extending the Basic Wavelet Framework

The basic wavelet framework presented in this book can be further developed in three main directions:

1. extending wavelets to general domains;
2. defining generalized representations using time-frequency decompositions, and
3. applying the wavelet framework to solve problems in applied mathematics.

We now briefly discuss what has been done in these three areas.

13.3.1 Studying Functions on other Domains

The first generalization is to study wavelets in other domains different than the real line.

It is natural to define wavelets in n-dimensional Euclidean spaces. The path to follow consists in extending the 1-D results that we have obtained to dimension $n > 1$. That is, to study wavelet transforms and multiresolution analyses on the space $\mathbf{L}^2(\mathbb{R}^n)$. We have several options in this direction. Some concepts and results extend naturally. There are two ways to construct wavelets in $\mathbb{R}^n$. The first way is based on a tensor product of one-dimensional wavelets. This extension is straightforward, and leads to a separable wavelet transform which can be implemented efficiently by multi-pass 1-D methods [20]. The second way is through a multidimensional multiresolution hierarchy generated by a single multivariate scaling function in $\mathbb{R}^n$ [50].

Another important extension is to define wavelets on compact subsets of the Euclidean space $\mathbb{R}^n$. A particular case concerns wavelets that are adapted to the unit interval $I = [0, 1]$. For this purpose, we have to construct a multiresolution of $\mathbf{L}^2(I)$, which is not generated from a single function anymore, but includes functions that adapt to the boundaries of the interval [15]. This construction extends naturally to rectangular regions of higher dimensions.

The final relevant extension should point us in the direction of defining wavelets and multiresolution analysis on arbitrary manifolds. Note that the methods we have described in this book make strong use of the Euclidean space structure, because of the translations used to obtain the wavelet basis. For this extension, we clearly need a methodology different from the Fourier Analysis. The Lifting scheme [57] and the similar construction in [17] are promising approaches in this direction.

13.3.2 Defining other Time-Frequency Decompositions

The second extension is to formulate function representations based on other time-frequency decomposition schemes.

Redundant wavelets bases are required for some operations. In particular, *dyadic wavelets* give a translation invariant representation that can be used for characterization of singularities and edge detection [39, 40]. Also, the *steerable pyramid* provides an orientation tuning mechanism [49].

Another generalization consists of decompositions based on different tilings of the time-frequency plane. Some examples are the wavelet packet basis [65], and the local cosine basis, also known as Malvar wavelets [41]. Some of these representations rely on a redundant dictionary of time-frequency atoms that can be used to construct descriptions adapted to individual functions. Best basis selection [16] and matching pursuit [38] are some of the optimization methods used to generate these representations. Similar techniques have been reported in [11] and [12].

A last generalization is a decomposition using a basis composed of several different functions, known as multiwavelets [41, 56].

13.3.3 Solving Mathematical Problems

The third direction is to study how wavelets can be applied in the solution of mathematical problems. This means that we will use wavelets to represent operators and to exploit the properties of this representation in order to derive more efficient computational schemes.

Wavelets have been used to solve integral equations [4], differential equations [2], and optimization problems. We remark that this last scheme has strong connections with multigrid methods.

13.4 Applications of Wavelets

Wavelets have been applied in many different areas. Here we will give an overview only of the applications in Computer Graphics and related areas.

We list below some of the main problems associated with the representation and processing of: geometric models, images, animation video sound and multimedia, that have been successfully solved using wavelets.

- **Signal and Image Processing**
  - Data compression
  - Progressive Transmission

  - Noise reduction
  - Filtering

- **Vision**
  - Edge detection
  - Texture Analysis
  - Feature Classification

- **Visualization and Graphics**
  - Radiosity
  - Volume Rendering
  - Paint Systems

- **Geometric Modeling**
  - Wavelets on surfaces and Variable resolution meshes
  - Multiscale Editing
  - Optimization of geometry
  - Synthesis of fractal surfaces

- **Animation**
  - Time-space constraints
  - Motion learning

- **Other Areas**
  - Medical Imaging
  - Geology
  - GIS and Cartography
  - Music
  - Speech synthesis and recognition
  - Databases for images and video.

A good review of wavelet application in computer graphics can be found in [47] and [51, 52]. A book dedicated entirely to this subject is [53].

Other sources of reference are the previous SIGGRAPH courses dedicated to wavelets [22, 48].

13.5 Comments and References

As a last remark, we mention that the Internet is a valuable source of information on current research and application of wavelets. From the many excellent websites, and other on-line resources, we would like to single out one: "The Wavelet Digest", an electronic newsletter moderated by Win Sweldens (http://www.wavelet.org/).

Appendix A
Systems and Filters

In this appendix we will review the basic concepts of Signal Processing to introduce the terminology and notation.

A.1 Systems and Filters

The word "system" has a very generic meaning in engineering. A system may change the input signal in different ways, producing an output signal.

A good example of a system is the human sound perception. It captures sounds, classifies and interprets them, in a way that we are not only able to understand the meaning, but we are also capable to identify the source of the sound. This complex task is accomplished by a long chain of processes. Among other things, the input sound is analyzed, separating its various frequency components using a set of filters with different sensibilities (this is performed by the cochlea, the basilar membrane, and other elements of the auditory system).

The term "filter" is employed to designate certain types of systems that alter only some frequencies of an input signal. The name has exactly this meaning: it implies that a selection is taking place and some frequency bands are altered, either attenuated or emphasized.

In the mathematical universe, a signal is modeled by a function. Therefore, we can model a system by a transformation $S : \mathscr{F}_1 \to \mathscr{F}_2$ between two function spaces, as illustrated in Fig. A.1.

The systems constitute the mathematical model to study the various operations involving processing of signals in the physical universe. We can give some examples:

J. Gomes, L. Velho, *From Fourier Analysis to Wavelets*,
IMPA Monographs 3, DOI 10.1007/978-3-319-22075-8

$$f \longrightarrow \boxed{\text{System}} \longrightarrow S(f) = g$$

Fig. A.1 Basic System

- The human eye is a system that processes electromagnetic signals in the visible range of the spectrum;
- A television camera has as input an electromagnetic signal (as the human eye), and its output is a video signal.

A.1.1 Spatial Invariant Linear Systems

In general, a system is characterized by some of its properties. Below, we describe the most relevant properties for our study:

Linearity

A system S is *linear* if:

1. $S(f + g) = S(f) + S(g)$
2. $S(\alpha f) = \alpha S(f)$

where f and g are two input functions (signals) and $\lambda \in \mathbb{R}$. In particular, $S(0) = 0$.

If a system S is linear and a signal f is represented by an atomic decomposition, for example:

$$f = \sum_{j=-\infty}^{\infty} \langle f, \varphi_j \rangle \varphi_j .$$

Its output $S(f)$ can be written as:

$$S(f) = \sum_{j=-\infty}^{\infty} \langle f, \varphi_j \rangle S(\varphi_j) .$$

That is, it is sufficient to know the output of the system for the basic decomposition atoms to predict the processing of any signal in this representation.

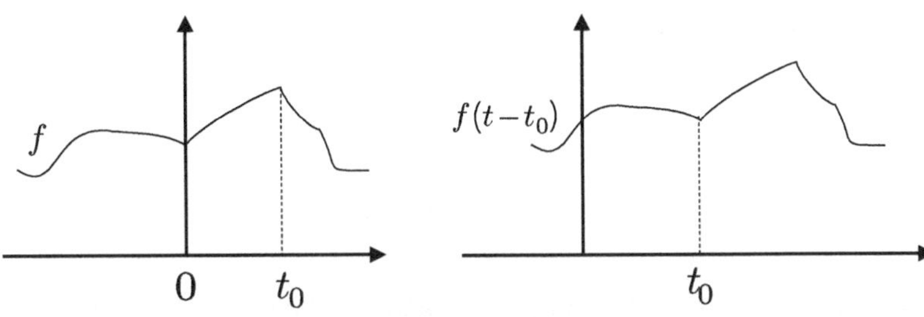

Fig. A.2 Delay Operation

Spatial Invariance

A system S is *spatial invariant* if it is possible to perform shifts in the signal, before or after its processing by the systems, and yet obtain the same result. That is,

$$S(f(t-t_0)) = S(f)(t-t_0)\,.$$

This class of system is generally called "time invariant" when the functions are of type $f : \mathbb{R} \to \mathbb{R}$, i.e. they have only one variable which usually represents the time evolution. When we work in higher dimensions, this time analogy is not valid anymore (even though time may still be one of the variables).

Given $t_0 \in \mathbb{R}$, the operator $R : \mathscr{F}_1 \to \mathscr{F}_2$ defined by

$$R(f) = f(t-t_0)$$

is called *delay operator by* t_0, or *translation operator by* t_0. Geometrically it makes a translation of t_0 units to the right in the graph of f, which is equivalent to a delay of t_0 time units in the signal (see Fig. A.2).

A system S is linear and spatial invariant if a delay in the input is equal to the same delay in the output

$$SR_{t_0} = R_{t_0}S\,.$$

Impulse Response

In this subsection we will characterize the spatial invariant linear systems.

The *impulse response*, h of a system S is the image $h = L(\delta)$ of the signal Dirac δ by the system.

If f is an arbitrary signal, we have:

$$f(x) = \int_{-\infty}^{\infty} f(t)\delta(x-t)dt\,.$$

If the system is spatial invariant, it follows from the above equation that:

$$S(f(x)) = \int_{-\infty}^{\infty} f(t)S(\delta(x-t))dt$$
$$= \int_{-\infty}^{\infty} f(t)h(x-t)dt\,.$$

The integral above is indicated by $f * h$, and is called *convolution product*. It is easy to verify that $f * h = h * f$. This result is summarized in the theorem below:

Theorem 8. *A spatial invariant linear system S is completely determined by its impulse response $h = S(\delta)$. More precisely, $S(f) = h * f$ for any signal f.*

The impulse response h gives a complete characterization of the system and, for this reason, it is called *filter kernel*.

A.1.2 Other Characteristics

For the moment, we will concentrate our attention only on these two properties mentioned above: linearity and time invariance. Nonetheless, there are many other important characteristics that deserve to be mentioned:

Finite Impulse Response

A system S has *finite impulse response*, (FIR), if its impulse response has compact support,

$$\text{supp}(S(\delta)) \subset [-L, L], \quad L < \infty\,.$$

The name originates from the fact that in the discrete domain, the impulse response of these filters is represented by a finite sequence.

Causality

Causality is a property which says that the output of the system depends only on previous facts in the input. That is, the system does not have any knowledge of the future.

A good example of a system in this class is the human ear, because sounds are processed strictly in chronological order.

Stability

Many stability conditions can be defined for a system, each of them related to the needs of different applications. A common stability condition is to enforce continuity of the operator which defines the system:

$$||S(f) \leq c||f|| \,.$$

In other words, if the input signal has finite energy, the same occurs with the output signal.

Taking the norm of sup: $||f|| = \sup(f)$, this condition reduces to:

$$\sup(S(f)) \leq c \sup(f) \,.$$

A system which satisfies this condition is called in the literature BIBO system ("Bounded Input–Bounded Output").

A.2 Discretization of Systems

We have to work with discrete systems to be able to implement the model of continuous systems in the computer. Therefore, we need to resort to methods to represent systems, which ultimately are methods to represent functions. Discrete systems operate with discrete signals.

A.2.1 Discrete Signals

As we have seen, various phenomena in the physical universe clearly have a discrete nature, such as turning a light on or off. Other phenomena, may even be conceived as continuous, but they are observed (measured) at discrete instants of time, for example, the variation of the tides.

Independently of a system being intrinsically continuous or discrete, in actual computer implementations we need to work with discrete representations of the system.

It is clear in my opinion a signal f we obtain a discretization that associates to the signal a sequence of "samples" (f_n), $n \in \mathbb{Z}$. We will use the notation $f[n], f(n)$ or even f_n to indicate the n-th "sample" of the signal. A sequence of samples of the signal will be indicated by (f_n). Sometimes, we will write $(f_n)_{n\in\mathbb{Z}}$ to emphasize that we are referring to a sequence.

The representation of a finite energy signal is defined by an operator $R\colon \mathscr{F} \to \ell^2$, of the space of signals into the space ℓ^2 of the square summable sequences: $R(f) =$

(f_n). The most common form of representation of a signal is given by its projection in a closed subspace $V \subset \mathscr{F}$. If $\{\varphi_j\}$ is an orthonormal base of this space, then

$$R(f) = \sum_k \langle f, \varphi_j \rangle \varphi_j \,.$$

Therefore, the samples of the signal are given by $f(j) = \langle f, \varphi_j \rangle$. Nonetheless, the reader could imagine that a sampling sequence is represented by uniform sampling the signal with rate Δt: $f(n) = f(n\Delta t)$ (we have seen that this is a good approximation for most cases of interest).

When the signal f is represented by a sequence (f_n), the description of the data elements which represent the physical measurements of the signal is done through the integer variable n. It is important to have a memory about the representation method employed, specially to make possible the reconstruction of the signal. In the case of point sampling, for example, it is important to know the sampling rate Δt.

If the representation sequence is finite, we have in fact a representation vector

$$(f_n) = (f(i), f(i+1), \ldots, f(j)) \,.$$

Otherwise, the representation sequence is an infinite vector

$$(f_n) = (\ldots, f(-2), f(-1), f(0), f(1), f(2), \ldots) \,.$$

A.2.2 Discrete Systems

Given a system $S : \mathscr{F}_1 \to \mathscr{F}_2$, we can take the representation $R_1 : \mathscr{F}_1 \to \ell^2$ and $R_2 : \mathscr{F}_2 \to \ell^2$, to produce the following diagram,

$$\begin{array}{ccc} \mathscr{F}_1 & \xrightarrow{S} & \mathscr{F}_2 \\ {\scriptstyle R_1}\downarrow & & \downarrow{\scriptstyle R_2} \\ \ell^2 & \xrightarrow{\overline{S}} & \ell^2 \end{array}$$

where $\overline{S}$ is the representation of the system S, that is, the discretized system S.

If S is linear and $\mathscr{F}_1$, $\mathscr{F}_2$ have a finite dimensional representation, that is, $R(\mathscr{F}_1) = \mathbb{R}^m$ and $R(\mathscr{F}_2) = \mathbb{R}^n$, then $\overline{S} : \mathbb{R}^m \to \mathbb{R}^n$ is a linear transformation, and can be represented by a matrix of order $n \times m$:

$$\begin{pmatrix} a_{11} & \cdots & a_{1m} \\ a_{21} & \cdots & a_{2m} \\ \vdots & \ddots & \vdots \\ a_{n1} & \cdots & a_{nm} \end{pmatrix} .$$

In case the representation of the spaces $\mathscr{F}_1, \mathscr{F}_2$ is not finite dimensional, we have $\overline{S}\colon \ell^2 \to \ell^2$. It is useful to represent these operators by infinite matrices

$$\begin{pmatrix} \vdots \\ y(-1) \\ y(0) \\ y(1) \\ \vdots \end{pmatrix} = \begin{pmatrix} \ddots & & & & \\ & a_{-1,-1} & a_{-1,0} & a_{-1,1} & \\ & a_{0,-1} & a_{0,0} & a_{0,1} & \\ & a_{1,-1} & a_{1,0} & a_{1,1} & \\ & & & & \ddots \end{pmatrix} \times \begin{pmatrix} \vdots \\ x(-1) \\ x(0) \\ x(1) \\ \vdots \end{pmatrix} .$$

If the system S is linear and spatially invariant, we know that

$$S(f) = h * f$$

where h is the kernel of the filter, $h = S(\delta)$. In the discrete domain $h = (h_n)$ and $f = (f_n)$, hence:

$$\begin{aligned} \overline{S}(f_n)(n) &= ((h_n) * (f_n))(n) \\ &= \sum_{k=-\infty}^{+\infty} h(k)f(n-k) \, , \end{aligned} \tag{A.1}$$

which is the direct expression for the convolution product.

Taking $(f_n) = (\delta_n)$, where (δ_n) is the discrete version of the Dirac delta:

$$(\delta_n)(k) = \begin{cases} 1 & \text{if } k = 0; \\ 0 & \text{if } k \neq 0. \end{cases}$$

then

$$S(\delta_n) = h * \delta_n = h \, ,$$

as it was expected.

The reader should observe that Eq. (A.1), which defines the output of a discrete linear system, is a finite difference linear equation:

$$y(n) = \sum_{k=-\infty}^{+\infty} h(k)f(n-k) \, ,$$

where $\overline{S}(f_n) = (y_n)$.

Because S is linear, we know that S is given by a matrix (possibly infinite). Now, we will obtain the matrix S using Eq. (A.1). For this, we need some linear algebra notation.

Indicating by R the unit delay operator: that is $R(f_n) = (f_{n-1})$ we have that

$$f(n-k) = R^k(f_n)$$

in other words, $f(n-k)$ is obtained from the signal (f_n) applying k delay operations. In particular, if $k = 0$, we have

$$f(n-0) = f(n) = R^0(f_n) = I(f_n)$$

where I is the identity operator. In this way, Eq. (A.1) can be written in operator form:

$$S(f_n) = \sum_{k=-\infty}^{+\infty} h(k)R^k(f_n) \ . \tag{A.2}$$

Given a matrix (a_{ij}), which can be infinite, the elements a_{jj} constitute the main diagonal.

$$\begin{pmatrix} \ddots & & & \\ & a_{0,0} & & \\ & & a_{1,1} & \\ & & & a_{2,2} \\ & & & & \ddots \end{pmatrix}$$

For each $d \in \mathbb{Z}$, $d > 0$, the elements $(a_{j+d,j})$ are the elements of the d-th lower diagonal.

$$\begin{pmatrix} \ddots & & & & \\ & 0 & & & \\ & 0 & 0 & & \\ & \bullet & 0 & 0 & \\ & 0 & \bullet & 0 & 0 \\ & & 0 & \bullet & 0 & 0 \\ & & & & & & \ddots \end{pmatrix}$$

Similarly, the elements $(a_{j,j+d})$ are the elements of the d-th upper diagonal.

The identity operator is represented by the identity matrix

$$I = \begin{cases} a_{j,j} = 1 \\ a_{i,j} = 0 \quad \text{if } i \neq j, \end{cases}$$

whose elements in the main diagonal are equal to 1, and the other elements are zero.

The unit delay operator

$$R(f_n) = f(n-1)$$

is represented by the matrix I_{-1} whose first lower diagonal consists of ones and the other elements are zero:

$$I_{-1} = \begin{cases} a_{j+1,j} = 1 & \\ a_{i,j} = 0 & \text{if } i \neq j+1. \end{cases}$$

$$\begin{pmatrix} \ddots & & & & & \\ & 0 & & & & \\ & 1 & 0 & & & \\ & 0 & 1 & 0 & & \\ & & 0 & 1 & 0 & \\ & & & & & \ddots \end{pmatrix} \tag{A.3}$$

More generally, the matrix of the operator R^k, shifted by k units, is:

$$I_{-k} = \begin{cases} a_{j+k,j} = 1 & \\ a_{i,j} = 0 & \text{if } i \neq j+k, \end{cases}$$

that is, all elements outside the k-th lower diagonal are zero, and the elements of this diagonal are equal to one.

Similar results hold for the operation

$$f(n) \longrightarrow f(n+k)$$

of time advance, considering the upper diagonals.

Observing Eq. (A.2) we see that the matrix of S is a matrix whose main, lower, and upper diagonals are constant. More precisely, the k-th diagonal is constituted by the elements $h(k)$.

$$\begin{pmatrix} \ddots & & & & & & \\ \cdots h(1) & h(0) & h(-1) & \cdots & & & \\ & \cdots h(1) & h(0) & h(-1) & \cdots & & \\ & & \cdots h(1) & h(0) & h(-1) & \cdots & \\ & & & \cdots h(1) & h(0) & h(-1) & \cdots \\ & & & & & & \ddots \end{pmatrix}$$

Observe that in the rows of this matrix we have the vector of the system impulse response, translated by one unit to the right from one line to the next. Because the output (y_n) will be the result of the inner product of the n-th line of the matrix by the input vector (x_n), we can either shift the input (x_n) before the multiplication or shift the output (y_n) after the multiplication: the result will be the same. This was expected, because the system represented by this matrix is linear and spatially invariant.

It is easy for the reader to verify that if the system is causal, all upper diagonals are zero (because the system cannot use advances of the signal in time). Therefore, the matrix is lower triangular.

Example 17. Consider the filter defined by the difference equation

$$y(n) = 3x(n-1) + 2x(n-2) + x(n-3) \ .$$

Then, we can rewrite this difference equation in the form of a convolution product

$$y(n) = \sum_k h(k)x(n-k)$$

where

$$h(k) = \begin{cases} 0 & \text{if } k \leq 0, \\ 3 & \text{if } k = 1, \\ 2 & \text{if } k = 2, \\ 1 & \text{if } k = 3, \\ 0 & \text{if } k \geq 4, \end{cases}$$

or, alternatively $(h_n) = (\ldots, 0, 0, 3, 2, 1, 0, 0, \ldots)$.

The matrix of this filter is given by

$$\begin{pmatrix} \vdots \\ \hat{x}(3) \\ \hat{x}(4) \\ \hat{x}(5) \\ \hat{x}(6) \\ \hat{x}(7) \\ \vdots \end{pmatrix} = \begin{pmatrix} \ddots & & & & & & \\ & 0 & & & & & \\ & 3 & 0 & & & & \\ & 2 & 3 & 0 & & & \\ & 1 & 2 & 3 & 0 & & \\ & 0 & 1 & 2 & 3 & 0 & \\ & & 0 & 1 & 2 & 3 & 0 \\ & & & & & & \ddots \end{pmatrix} \times \begin{pmatrix} \vdots \\ x(0) \\ x(1) \\ x(2) \\ x(3) \\ x(4) \\ x(5) \\ \vdots \end{pmatrix} .$$

Note that the filter is causal

Question A.1. What would happen if the coefficients of the difference equation were not constant?

The equation would be described as

$$y_n = \cdots + h_0(n)x_n + h_1(n)x_{n-1} + h_2(n)x \, .$$

In this case the filter kernel (h_n) is a sequence which varies with the index (time), therefore, it is necessary now to represent it as:

$$(h_n) = (\ldots, h_{-1}(n), h_0(n), h_1(n), \ldots) \, .$$

The convolution product cannot be used to characterize the operation and its "generalization" is

$$(y_n) = \sum_k h_k(n)x_{n-k} \, .$$

Finally, the matrix representation does not have constant diagonals:

$$\begin{pmatrix} \vdots \\ y_{-1} \\ y_0 \\ y_1 \\ \vdots \end{pmatrix} = \begin{pmatrix} \ddots & & & & \\ \cdots & h_1(-1) & h_0(-1) & h_{-1}(-1) & \cdots \\ \cdots & h_2(0) & h_1(0) & h_0(0) & \cdots \\ \cdots & h_3(1) & h_2(1) & h_1(1) & \cdots \\ & & & & \ddots \end{pmatrix} \times \begin{pmatrix} \vdots \\ x_{-1} \\ x_0 \\ x_1 \\ \vdots \end{pmatrix} .$$

A.3 Upsampling and Downsampling Operators

Besides the delay operator, two other operators are important in the study of discrete filters: they are the downsampling and upsampling operators.

The *downsampling* operator of order q, $\downarrow \mathrm{q}\colon \ell^2 \to \ell^2$, is defined as

$$(\downarrow \mathrm{q})(u_n) = (u_{nq}) \, .$$

That is, the operator discards all terms of a sequence, except the terms which are multiple of q. This operator is also known in the literature as the *decimation* operator of order q.

Here, we will only study the case where $q = 2$. In this case the operator discards alternating terms in the sequence, retaining only the terms with even index:

$$(\downarrow 2)(u_n) = (u_{2n}) \, .$$

The matrix of this operator is given by

$$\begin{pmatrix} 1 & & & & & \\ 0 & 0 & 1 & & & \\ & & 0 & 0 & 1 & \\ & & & & 0 & \\ & & & & & \cdots \end{pmatrix}.$$

This matrix is obtained from the identity matrix I, by including a column of zeros in alternation. Or equivalently, we can shift the lines of the matrix one unit to the right in alternation.

It is immediate to verify that the downsampling operator is not invertible. However, it has an inverse to the right, which is the *upsampling* operator:

$$(\uparrow 2)(\ldots, x_{-1}, x_0, x_1, \ldots) = (\ldots, x_{-1}, 0, x_0, 0, x_1, 0, \ldots) .$$

That is,

$$(\uparrow 2)(u_n)(k) = \begin{cases} u(k) & \text{if } n = 2k \\ 0 & \text{if } n = 2k+1 . \end{cases}$$

The upsampling operator simply intercalates zeros in between the elements of the representation sequence.[1]

It is easy for the reader to verify that the upsampling operator $\uparrow 2$ is the inverse to the right downsampling operator $\downarrow 2$, that is $(\downarrow 2)(\uparrow 2) = I$. However, $(\uparrow 2)(\downarrow 2) \neq I$, as we can see below:

$$(\uparrow 2)(\downarrow 2)\begin{pmatrix} \vdots \\ u_{-1} \\ u_0 \\ u_1 \\ \vdots \end{pmatrix} = \begin{pmatrix} \vdots \\ u_{-2} \\ 0 \\ u_0 \\ 0 \\ u_2 \\ 0 \\ \vdots \end{pmatrix}.$$

In terms of matrices, it is easy to see that the matrix of the operator $\uparrow 2$ is obtained by intercalating rows of zeros in the identity matrix

[1]We can define the upsampling operator of order q, intercalating q zeros.

$$\uparrow 2 = \begin{pmatrix} 1\,0 & & & \\ & 0\,0 & & \\ & 1\,0\,0 & & \\ & & 0\,0 & \\ & & 1\,0 & \\ & & & \dots \end{pmatrix}.$$

This is equivalent to translate the columns of the identity matrix by one unit to the bottom.

It is immediate to verify that the matrix of the upsampling operator is the transpose of the matrix of the downsampling operator, and vice versa. The relation between the operators of downsampling and upsampling can be stated using matrices:

$$(\downarrow 2)(\uparrow 2) = I,$$

and

$$(\uparrow 2)(\downarrow 2) = \begin{pmatrix} 1 & & & & & \\ & 0 & & & & \\ & & 1 & & & \\ & & & 0 & & \\ & & & & 1 & \\ & & & & & \ddots \end{pmatrix}.$$

Once more, we observe that the upsampling operator is the inverse to the right, but is not an inverse to the left. At this point we can argue that the nature of the mathematical objects is against us: as we will see, the most important operation is to recover a signal after downsampling, and not the other way around.

A.4 Filter Banks

A *filter bank* is a system composed of several filters, together with delay, upsampling and downsampling operators.

A filter bank with two filters, downsampling, and upsampling is illustrated in Fig. A.3.

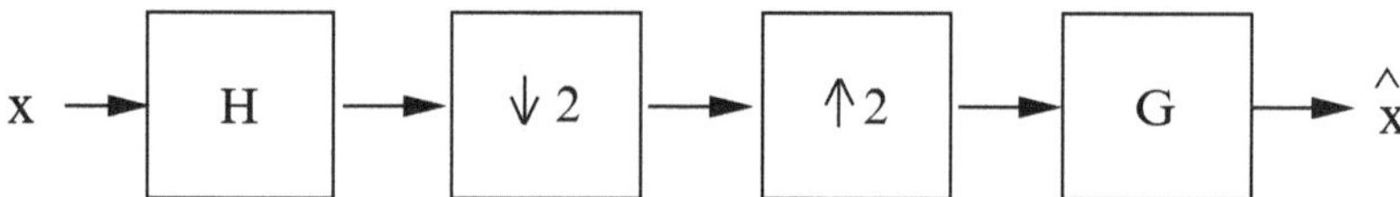

Fig. A.3 Example of a Filter Bank

A.5 Comments and References

A basic reference for signal processing is [44]. It includes the whole theory of signals and systems, from the continuous to the discrete. Another good source, but somewhat older, discussing only continuous systems is [32].

For a book more applied to the project and analysis of filters, the reader can use [1].

The literature of signal processing has several sources for the reader interested in systems with multirate filter banks, one of them is [61].

Appendix B
The Z Transform

The behavior of filters in the frequency domain gives a good intuition of some of its characteristics, which are difficult to grasp by looking only at the filter coefficients in the time domain. We have seen that the Fourier transform is a tool used to study a signal in the frequency domain. In this appendix, we will introduce a similar, but more general transform, the Z transform, which in fact contains the Fourier transform.

B.1 The Z Transform

Given a linear, time invariant, discrete system S with impulse response (h_n), and an input signal (x_n), the output is then $(y_n) = (h_n) * (x_n)$. However, let's choose an input signal with a special structure:

$$x(n) = z^n , \qquad z \in \mathbb{Z}$$

in this case, the output signal is then:

$$\begin{aligned}(y_n) = (h_n) * (x_n) &= \sum_{k=-\infty}^{+\infty} h(k)x(n-k) \\ &= \sum_{k=-\infty}^{+\infty} h(k)z^{n-k} = z^n \sum_{k=-\infty}^{+\infty} h(k)z^{-k}\end{aligned}$$

J. Gomes, L. Velho, *From Fourier Analysis to Wavelets*,
IMPA Monographs 3, DOI 10.1007/978-3-319-22075-8

that is, the value of the output signal $y(n)$ is obtained by multiplying $x(n)$ by a constant. This constant, however, varies with the value of z, which was fixed for each input signal:

$$y(n) = H(z)z^n$$

where

$$H(z) = \sum_{k=-\infty}^{+\infty} h(k)z^{-k} .$$

We should remark that if an arbitrary signal is decomposed as a direct sum of basic atoms, it would be sufficient to know the effect of the system over its atoms to be able to predict the result of the system on any input signal. For this reason, we have chosen our input signal as a sum of different complex exponentials:

$$x'(n) = \sum_k a_k z_k^n$$

and, applying it to the system S, we can use its linearity property to obtain:

$$y'(n) = \sum_k a_k H(z_k) z_k^n .$$

Using a linear algebra terminology, the complex exponentials z_k^n are *eigenvector functions* of S, and $H(z_k)$ are their respective *eigenvalues*.

Exploiting the above intuition, we will introduce the $\mathscr{Z}$ transform.

Given a discrete signal (x_n), its $\mathscr{Z}$ transform is given by:

$$X(z) = \mathscr{Z}\{x(n)\} = \sum_{k=-\infty}^{\infty} z^{-k} x(k) . \tag{B.1}$$

Together with $X(z)$ it is necessary to know the corresponding *Convergence Region*, that is, the regions of the complex plane in which the variable z is defined, such that this transform converges.

Remark B.1. Based on this transform, making $z = e^{i2\pi w}$, we obtain the Fourier transform for discrete aperiodic signals (DTFT – "Discrete Time Fourier Transform").

B.1.1 Some Properties

Several important properties of the $\mathscr{Z}$ transform will help in the manipulation of discrete systems. We will see that these properties contribute to the intuitive view alluded to in the beginning of this Appendix.

- Linearity;

$$\mathscr{Z}\{x_1(n) + x_2(n)\} = \mathscr{Z}\{x_1(n)\} + \mathscr{Z}\{x_2(n)\}$$

- Time Shift;

$$\mathscr{Z}\{x(n-l)\} = \sum_{k=-\infty}^{\infty} z^{-k}x(k-l) = \sum_{k=-\infty}^{\infty} z^{-k+l}x(k)$$
$$= z^{l}\mathscr{Z}\{x(n)\}$$

- Time Reversal;

$$x(n) \leftrightarrow X(z) \qquad \longleftrightarrow \qquad x(-n) \leftrightarrow X(z^{-1})$$

  This property, as the next one, is easily verified, just by applying the definition of the transform.
- Convolution in Time is Multiplication in Frequency;

$$\mathscr{Z}\{x_1(n) * x_2(n)\} = X_1(z)X_2(z)$$

Remark B.2. The function delta, δ, exhibits a special behavior. We have seen when it was the input of a system, the output was the impulse response function, which characterizes the filter. This has two main reasons: First, the convolution product in the time domain is equivalent to multiplication in frequency; Second, the representation of the delta function in the frequency domain is:

$$\mathscr{Z}\{\delta(n)\} = 1 \ .$$

This means that the function delta $\delta(n)$ is the *neutral element* of the convolution operation in time and of the multiplication operation in frequency.

B.1.2 Transfer Function

The impulse response $h(t)$ of a filter has a correspondent in the frequency domain, which is called *Transfer Function*.

Given a system S, its impulse response $h = S(\delta)$ completely defines the filtering operation

$$S(f) = h * f \; .$$

What would be the interpretation of the fact above in the frequency domain, using the $\mathscr{Z}$ Transform?

Suppose that $S(x_n) = (y_n)$, that is,

$$(y_n) = (h_n) * (x_n) \; .$$

Going to the frequency domain, and using the properties of the $\mathscr{Z}$ transform, we have:

$$Y(z) = \mathscr{Z}\{(h_n) * (x_n)\} = \mathscr{Z}\{(h_n)\}.\mathscr{Z}\{(x_n)\} = H(z)X(z) \; .$$

We isolate $H(z) = \mathscr{Z}\{(h_n)\}$, and call this expression *Transfer Function*:

$$H(z) = \frac{Y(z)}{X(z)} \; .$$

Again, it is necessary to be careful with the convergence of this expression, and the division operation takes care that the zeros of $X(z)$ are outside the convergence region.

The transfer function $H(z)$ gives exactly information on how the input frequencies (x_n) will be altered in order to produce the output (y_n). This connection was one of the goals of our intuitive view. A linear, time invariant, discrete system will react in a predictable manner according to the complex exponentials applied to its input. Its output will be given as the same input exponential, but with a different "amplitude."

Considering that a filter is exactly the system that makes this type of transformation, then the $\mathscr{Z}$ transform is a good way to describe filters. What would be the disadvantages of using (h_n)?

One is immediate: the variable z is not discrete, it is also defined in the entire complex plane (in the Convergence Region). To store it in the computer it is necessary to do some kind of "sampling." The other option would be the use of a symbolic machinery, but the manipulation would be more difficult and less efficient.

For the tasks of filter design and evaluation, the importance of the $\mathscr{Z}$ transform is undisputable. To obtain the transfer function $H(z)$ from the linear difference equation with constant coefficients, i.e. the convolution product, is a direct operation: the coefficients of the equation are exactly the coefficients of a polynomial in z:

$$\mathscr{Z}\{y(n)\} = \mathscr{Z}\left\{\sum_k b_k x(n-k)\right\}$$

$$Y(z) = \left(\sum_k b_k z^{-k}\right) X(z)$$

and then,

$$H(z) = \sum_k b_k z^{-k} .$$

In a more general form, we can have:

$$\mathscr{Z}\left\{\sum_{k=0}^{N} a_k y(n-k)\right\} = \mathscr{Z}\left\{\sum_k b_k x(n-k)\right\}$$

$$\left(\sum_{k=0}^{N} a_k z^- k\right) Y(z) = \left(\sum_k b_k z^{-k}\right) X(z)$$

in this case $H(z)$ is a ratio of polynomials

$$H(z) = \frac{\sum\limits_k b_k z^{-k}}{\sum\limits_{k=0}^{N} a_k z^- k} .$$

Example 18. Going back to the system of example 17, in the previous chapter, it will be instructive to determine the system transfer function, and look at its frequency behavior

$$\hat{x}(n) = 3x(n-1) + 2x(n-2) + x(n-3)$$
$$\mathscr{Z}\{y(n)\} = \mathscr{Z}\{3x(n-1) + 2x(n-2) + x(n-3)\} .$$

Exploiting the linearity and time shift properties, we have:

$$Y(z) = 3z^{-1}X(z) + 2z^{-2}X(z) + z^{-3}X(z)$$

and factoring out $X(z)$,

$$H(z) = \frac{Y(z)}{X(z)} = 3z^{-1} + 2z^{-2} + z^{-3} .$$

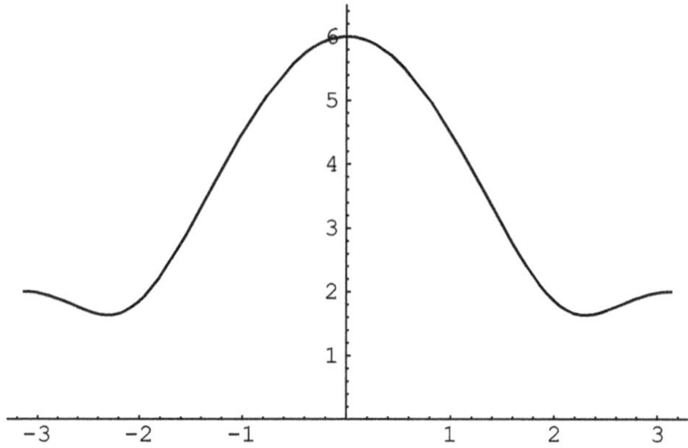

Fig. B.1 Modulus of the frequency response of a filter

In order to really visualize the behavior of a filter in the different frequencies that compose the input signal, it is sufficient to evaluate the transfer function $H(z)$ for the values of $z = e^{iw}$, where $w = [0, \pi]$. In this last example $|H(e^{j}w)|$ is shown in Fig. B.1.

B.1.3 The Variable z and Frequency

Writing the complex variable z in polar form, we obtain

$$z = \alpha e^{iw}, \qquad \text{where} \quad 0 < w < \pi$$

The discrete domain has a limited interval of possible frequencies due to the sampling process. Making a normalization, all these possible frequencies will be contained in the interval $[0, \pi]$, where π has a direct relation with $1/2$ of the sampling frequency used in the discretization process.

Therefore, it makes sense to consider the function $H(e^{iw})$ as a periodic function. Again, $H(e^{iw})$ is also the DTFT of the non-periodic filter (h_{n}) and every DTFT is periodic with a period of 2π.

Through the behavior of systems in the various frequency regions, it is possible also to classify the filters into basic types, such as low-pass, high-pass, band-pass, etc., as we have seen in Chap. 3.

B.2 Subsampling Operations

In this section we will discuss the downsampling and upsampling operators in the frequency domain. We will derive the expressions for these operators using the z-notation.

B.2.1 Downsampling in the Frequency Domain

The downsampling operator, introduced in the previous chapter, takes a discrete sequence $(x_k)_{k\in\mathbb{Z}}$ and removes all elements with odd indices, keeping only even index elements. Therefore, $v = (\downarrow 2)x$ implies that $v_k = x_{2k}$.

In order to derive the expression for the downsampling operator in the frequency domain, we will take a sequence (u_n), which has all the elements with even index from an arbitrary sequence (x_n), and the elements with odd indices equal to zero

$$u_n = \begin{cases} x_n & \text{if } n \text{ is even} \\ 0 & \text{if } n \text{ is odd} . \end{cases}$$

That is, $u = (\dots, x_0, 0, x_2, \dots)$. Clearly, $(\downarrow 2)x = (\downarrow 2)u$. So, let's write the Fourier transform of u

$$\begin{aligned} U(\omega) &= \sum_{n \text{ even}} x_n e^{-in\omega} \\ &= \frac{1}{2}\sum_{\text{all } n} x_n e^{-in\omega} + \frac{1}{2}\sum_{\text{all } n} x_n e^{-in(\omega+\pi)} . \end{aligned} \tag{B.2}$$

The expression for U was split into two terms, so that when they are added together only the terms with even indices remain. This is because, if $n = 2l$ is even, then $e^{-i(2l)(\omega+\pi)} = e^{-i2l\omega+2l\pi} = e^{-i2l\omega}$, preserving the even index elements. But, if $n = 2l + 1$ is odd, then $e^{-i(2l+1)(\omega+\pi)} = e^{-i(2l+1)\omega+(2l+1)\pi} = e^{-i(2l+1)\omega+\pi} = -e^{-i2l\omega}$, removing the odd index elements.

Therefore, the expression of $U(\omega)$ in the frequency domain can be written in terms of $X(\omega)$ as $U(\omega) = \frac{1}{2}[X(\omega) + X(\omega + \pi)]$.

Now, using $U(\omega)$ which contains only even index terms, we can write the formula for the downsampling operator in frequency domain just by halving the frequencies $V(\omega) = U(\omega/2)$.

$$V(\omega) = \frac{1}{2}\left[X\left(\frac{\omega}{2}\right) + X\left(\frac{\omega}{2} + \pi\right)\right] . \tag{B.3}$$

The downsampling operation corresponds to a change in the sampling rate, as it is implied by halving the frequencies $\omega \to \omega/2$. This may cause *aliasing*, which

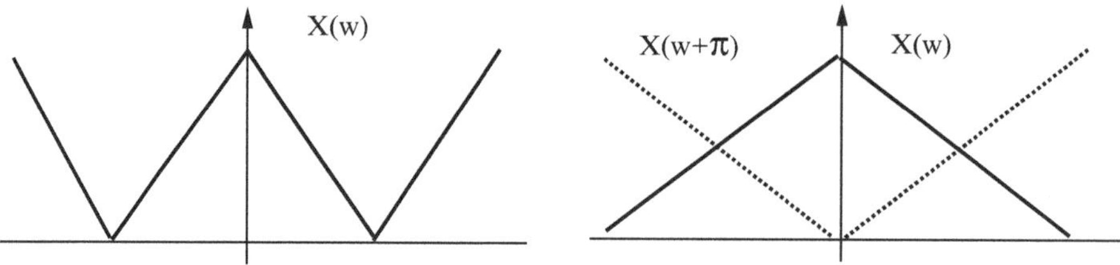

Fig. B.2 Downsampling Operator

is essentially a frequency folding, manifested by the introduction of a new term in $\omega/2 + \pi$. This process is illustrated in Fig. B.2.

In the $\mathscr{Z}$-domain, the downsampling operator is

$$V(z) = \frac{1}{2}\left[X(z^{1/2}) + X(-z^{1/2})\right] . \tag{B.4}$$

B.2.2 Upsampling in the Frequency Domain

The upsampling operator, also introduced in the previous chapter, takes a discrete sequence $(x_k)_{k\in\mathbb{Z}}$ and an interleave zeros in between the sequence elements.

$$u = (\uparrow 2)v \Leftrightarrow \begin{cases} u_{2k} & = v_k \\ u_{2k+1} & = 0 \end{cases}$$

In the frequency domain, this operation has a simple expression. We only retain the terms with even index $n = 2k$, because terms with odd index are zero, $u_{2k+1} = 0$.

$$\begin{aligned} U(\omega) &= \sum u_n e^{-in\omega} = \sum u_{2k} e^{-i2k\omega} \\ &= \sum v_k e^{-i2k\omega} \end{aligned} \tag{B.5}$$

So, $u = (\uparrow 2)v$ is $U(\omega) = V(2\omega)$, or $U(z) = V(z^2)$.

The upsampling operation causes *imaging*. A bandlimited function $V(\omega)$ with period 2π is mapped into a function $U(\omega) = V(2\omega)$ with period π. Compressed copies of the graph $V(2\omega)$ appear in the spectrum of $U(\omega)$. This is illustrated in Fig. B.3.

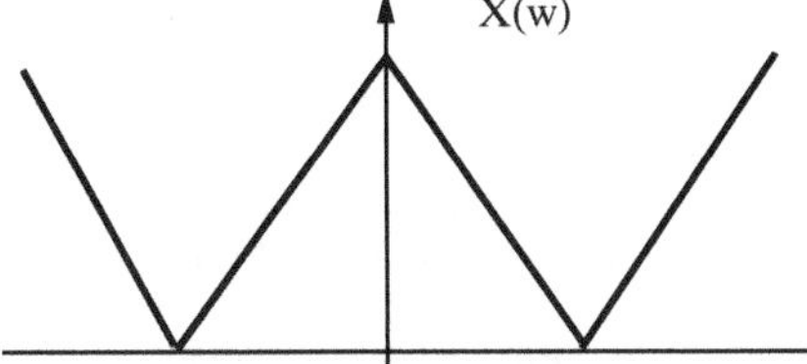

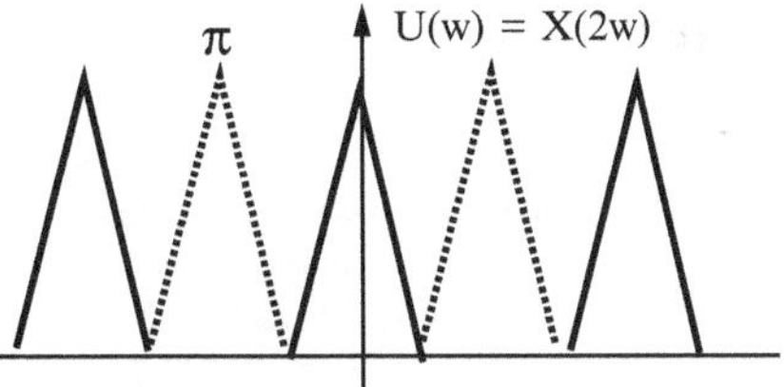

Fig. B.3 Upsampling Operator

B.2.3 Upsampling after Downsampling

In a filter bank it is common to apply both the downsampling and upsampling operations. This is because, usually downsampling is part of the analysis bank, and upsampling is part of the synthesis bank.

Let's see how these two operations can be combined together:

$$v = (\downarrow 2)x \quad \text{and} \quad u = (\uparrow 2)v$$

this result is $u = (\uparrow 2)(\downarrow 2)x$, which in the frequency domain is

$$U(\omega) = \frac{1}{2}[X(\omega) + X(\omega + \pi)] \tag{B.6}$$

or in the Z-domain

$$U(z) = \frac{1}{2}[X(z) + X(-z)] \,. \tag{B.7}$$

B.3 Comments and References

A good reference for the use of filter banks and their relations with wavelets is [55]. A more general book about multirate systems and filter banks is [61].

References

1. Andreas Antoniou. *Digital Filters*. McGraw-Hill, second edition, 1993.
2. Emmanual Bacry, Stephane Mallat, and George Papanicolaou. A wavelet based space–time adaptive numerical method for partial differential equations. *Mathematical Modelling and Numerical Analysis*, 26:793–834, 1992.
3. G. Beylkin, R. Coifman, and V. Rokhlin. Fast wavelet transforms and numerical algorithms. *Comm. in Pure and Applied Math.*, 44:141–183, 1991.
4. G. Beylkin, R. Coifman, and V. Rokhlin. Fast wavelet transforms and numerical algorithms I. *Communications on Pure and Applied Mathematics*, XLIV:141–183, 1991. transformation en ondelettes.
5. L. Blum. *Lectures on a Theory of Computation and Complexity over the Reals*. 18 Coloquio Brasileiro de Matematica – IMPA, 1991.
6. Mikael Bourges-Sévenier. Réalisation d'une bibliothèque c de fonctions ondelettes. Technical report, IRISA – INRIA, 1994.
7. W. L. Briggs and V. E. Henson. *DFT: an owner's manual for the discrete Fourier transform*. SIAM books, Philadelphia, 1995.
8. P. J. Burt and E. H. Adelson. The Laplacian pyramid as a compact image code. *IEEE Trans. Commun.*, 31(4):532–540, April 1983.
9. P. L. Butzer and R. L. Stens. Sampling theory for not necessarily band-limited functions: a historical overview. *SIAM Reviews*, 34(1):40–53, march 1992.
10. Claudio Canuto and Anita Tabacco. Multilevel decompositions of functional spaces. *The Journal of Fourier Analysis and Applications*, 3(6), 1997.
11. Shaobing Chen and David L. Donoho. Basis pursuit. Technical report, Stanford University, 1994.
12. Shaobing Chen and David L. Donoho. Atomic decomposition by basis pursuit. Technical report, Stanford University, 1995.
13. C. K. Chui. *An introduction to wavelets*. Academic Press, 1992.
14. A. Cohen, I. Daubechies, and J. C. Feauveau. Biorthogonal bases of compactly supported wavelets. *Comm. Pure Applied Math.*, 1992.
15. A. Cohen, I. Daubechies, and P. Vial. Wavelets on the interval and fast wavelet transforms. *Applied and Computational Harmonic Analysis*, 1(1):54–81, 1993.
16. Ronald R. Coifman and Mladen Victor Wickerhauser. Entropy-based algorithms for best basis selection. *IEEE Transactions on Information Theory*, 38(2):713–718, 1992.
17. Wolfgang Dahmen and Reinhold Schneider. Wavelets on manifolds i: Construction and domain decomposition. Technical report, Institut fur Geometrie und Praktische Mathematik, 1998.
18. I. Daubechies. Orthonormal bases of compactly supported wavelets. *Comm. Pure Applied Math.*, XLI(41):909–996, November 1988.
19. I. Daubechies, A. Grossmann, and Y. Meyer. Painless non-orthogonal expansions. *J. Math. Phys.*, (27):1271–1283, 1986.
20. Ingrid Daubechies. *Ten Lectures on Wavelets*. SIAM Books, Philadelphia, PA, 1992.
21. P. Duhamel and M. Vetterli. Fast fourier transform: A tutorial review and a state of the art. *Signal Proc.*, 19(4):259–299, 1990.
22. Alain Fournier. Wavelets and their applications in computer graphics. SIGGRAPH Course Notes, http://www.cs.ubc.ca/nest/imager/contributions/bobl/wvlt/download/notes.ps.Z.saveme, 1994.
23. D. Gabor. Theory of communication. *J. of the IEEE*, (93):429–457, 1946.
24. J. Gomes and Luiz Velho. *Image Processing for Computer Graphics*. Springer-Verlag, 1997.
25. Jonas Gomes, Bruno Costa, Lucia Darsa, and Luiz Velho. Graphical objects. *The Visual Computer*, 12:269–282, 1996.
26. Jonas Gomes and Luiz Velho. Abstraction paradigms for computer graphics. *The Visual Computer*, 11:227–239, 1995.

27. Donald Greenspan. *Computer Oriented Mathematical Physics*. International Series in Nonlinear Mathematics, methods and Applications. Pergamon Press, New York., 1981.
28. E. Hernandez and G. Weiss. *A First Course on Wavelets*. CRC Press, Boca Raton, 1996.
29. Nicholas J. Higham. *Accuracy and Stability of Numerical Algorithms*. SIAM Books, Philadelphia, 1996.
30. B. Jawerth and Wim Sweldens. An overview of wavelet based multiresolution analyses. *SIAM Rev.*, 36(3):377–412, 1994.
31. G. Kaiser. *A Friendly Guide to Wavelets*. Birkhauser, Boston, 1994.
32. Bhagwandas P. Lathi. *Signals, Systems, and Control*. Harper & Row, 1974.
33. Jae S. Lim. *Two Dimensional Signal and Image Processing*. Prentice-Hall, New York, 1990.
34. Van Loan. *Computational Frameworks for the Fast Fourier Transform*. SIAM books, Philadelphia, 1996.
35. S. Mallat. Multifrequency channel decomposition of images and wavelet models. *IEEE Transaction on ASSP*, 37:2091–2110, 1989.
36. S. Mallat. Multiresolution approximation and wavelets. *Trans. Amer. Math. Soc.*, 315:69–88, 1989.
37. S. Mallat. *A Wavelet Tour of Signal Processing*. Academic Press, 1998.
38. S. Mallat and A. Zhang. Matching pursuit with time-frequency dictionaries. Technical report, Courant Institute of Mathematical Sciences, 1993.
39. S. Mallat and S. Zhong. Characterization of signals from multiscale edges. *IEEE Transactions on Pattern Analysis and Machine Intelligence*, 14(7):710–732, 1992.
40. S. G. Mallat and W. L. Hwang. Singularity detection and processing with wavelets. *IEEE Transactions on Information Theory*, 38(2):617–643, 1992.
41. H. S. Malvar. Modulated QMF filter banks with perfect reconstruction. *Electronics Letters*, 26:906–907, June 1990.
42. A. I. Markushevich. *Theory of Functions of a Complex Variable*. Chelsea Publishing Co., 1977.
43. Y. Meyer. Construction de bases orthonormées d'ondelettes. *Colloq. Math.*, 60/61:141–149, 1990. bib.
44. Alan V. Oppenhein and Alan S. Willsky. *Signal and Systems*. Prentice-Hall inc, 1983.
45. William H. Press, Saul A. Teukolsky, and William T. Vetterling. *Numerical Recipes : The Art of Scientific Computing*, chapter 13, pages 591–606. Cambridge Univ Press, 1996.
46. A. A. G. Requicha. Representations for rigid solids: Theory methods, and systems. *ACM Computing Surveys*, 12:437–464, December 1980.
47. Peter Schroder. Wavelets in computer graphics. In *Proceedings of the IEEE*, volume 84, pages 615–625, 1996.
48. Peter Schroder and Wim Sweldens. Wavelets in computer graphics. SIGGRAPH Course Notes, http://www.multires.caltech.edu/teaching/courses/waveletcourse/, 1995.
49. Eero P. Simoncelli and William T. Freeman. The steerable pyramid: A flexible architecture for multi-scale derivative computation. In *International Conference on Image Processing*, volume 3, pages 444–447, 23–26 Oct. 1995, Washington, DC, USA, October 1995.
50. J. Stöckler. Multivariate wavelets. In C. K. Chui, editor, *Wavelets: A Tutorial in Theory and Applications*, pages 325–356. Academic Press, 1992.
51. Eric J. Stollnitz, Tony D. DeRose, and David H. Salesin. Wavelets for computer graphics: a primer, part 1. *IEEE Computer Graphics and Applications*, 15(3):76–84, May 1995.
52. Eric J. Stollnitz, Tony D. DeRose, and David H. Salesin. Wavelets for computer graphics: a primer, part 2. *IEEE Computer Graphics and Applications*, 15(4):75–85, July 1995.
53. Eric J. Stollnitz, Tony D. DeRose, and David H. Salesin. *Wavelets for Computer Graphics: Theory and Applications*. Morgann Kaufmann, San Francisco, CA, 1996.
54. G. Strang and J. Shen. The zeros of the Daubechies polynomials. *Proc. Amer. Math. Soc.)*, 1996.
55. Gilbert Strang and Truong Nguyen. *Wavelets and Filter Banks*. Wellesley-Cambridge Press, Wellesley, MA, 1996.

56. Gilbert Strang and Vasily Strela. Orthogonal multiwavelets with vanishing moments. *Optical Engineering*, 33(7):2104–2107, 1994.
57. W. Sweldens. The lifting scheme: A construction of second generation wavelets. Technical Report 1995:6, Department of Mathematics, University of South Carolina, 1995.
58. Carl Taswell. Algorithms for the generation of Daubechies orthogonal least asymmetric wavelets and the computation of their Holder regularity. Technical report, Scientific Computing and Computational Mathematics, Stanford University, Stanford, CA, August 1995.
59. Carl Taswell. *Computational Algorithms for Daubechies Least-Asymmetric, Symmetric, and Most-Symmetric Wavelets*, pages 1834–1838. Miller Freeman, September 1997.
60. Carl Taswell. The systematized collection of wavelet filters computable by spectral factorization of the Daubechies polynomial. Technical report, Computational Toolsmiths, www.toolsmiths.com, December 1997.
61. P. P. Vaidyanathan. *Multirate Systems and Filter Banks*. Prentice Hall PTR, Englewood Cliffs, New Jersey, 1993.
62. M. Vetterli and C. Herley. Wavelets and filter banks: theory and design. *IEEE Trans. Acoust. Speech Signal Process.*, 40(9), 1992.
63. Martin Vetterli and Jelena Kovacevic. *Wavelets and Subband Coding*. Prentice Hall PTR, Englewood Cliffs, New Jersey, 1995.
64. J. Weaver. *Theory of Discrete and Continuous Fourier Transform*. John Wiley & Sons, New York, 1989.
65. Mladen Victor Wickerhauser. *Adapted Wavelet Analysis from Theory to Software*. A. K. Peters, Wellesley, MA, 1994.
66. Ahmed Zayed. *Advances in Shannon's Sampling Theory*. CRC Press, Boca Raton, 1993.

Zeitfracht Medien GmbH
Ferdinand-Jühlke-Straße 7
99095 Erfurt, Deutschland
produktsicherheit@kolibri360.de